Natural Resou  n
Wild ...s

Natural Resource Administration
Wildlife, Fisheries, Forests and Parks

Donald W. Sparling

ELSEVIER

AMSTERDAM • BOSTON • HEIDELBERG • LONDON
NEW YORK • OXFORD • PARIS • SAN DIEGO
SAN FRANCISCO • SINGAPORE • SYDNEY • TOKYO
Academic Press is an imprint of Elsevier

Academic Press is an imprint of Elsevier
525 B Street, Suite 1800, San Diego, CA 92101-4495, USA
32 Jamestown Road, London NW1 7BY, UK
225 Wyman Street, Waltham, MA 02451, USA

Notice

British Library Cataloguing-in-Publication Data
A catalogue record for this book is available from the British Library

Library of Congress Cataloging-in-Publication Data
A catalog record for this book is available from the Library of Congress

ISBN: 978-0-12-404647-4

For information on all Academic Press publications
visit our website at elsevierdirect.com

Typeset by MPS Limited, Chennai, India
www.adi-mps.com

Working together
to grow libraries in
developing countries

www.elsevier.com • www.bookaid.org

Dedication

There are many people I am indebted to for helping with this book. There are the thousands of recognized and unrecognized field personnel, scientists, administrators, legislators, judges, and NGO personnel that have in some way helped shape the organization that we now have supporting conservation efforts in the United States and Canada. This book is dedicated to you.

Most of all, I would like to thank my wife, Paulette, for her understanding and care.

Contents

Part III
The Bureaucracy of Natural Resources

Part IV
Non-Governmental Agencies, People, and Money

Introduction

Over the years of teaching classes in wildlife administration, I have found the reasons why students take the course to be very interesting. When I was an undergraduate, and even as a graduate student in zoology a bunch of years ago, I was much more interested in taking what I referred to as the "ologies" – herpetology, ornithology, mammalogy, wildlife biology, or zoology – and learning what types of animals existed, how they lived, communicated and reproduced, and how they could be conserved, than a course that taught about bureaucracy, laws, public relations, budgets, and other related (and, as I thought then, esoteric) topics. After earning my doctorate I wanted a position in academia, but the market was extremely tight – in one year only three positions in my area were advertised, and 360 other PhDs were applying for them. So, my first permanent career position was in the US Fish and Wildlife Service. Immediately I was inundated with much of the information presented in this book, and the learning curve was steep. One lesson I received in those days was that a person has to be flexible and ready to take a good opportunity when it is presented. Another lesson I learned over the years was that working for a federal, state, or provincial agency can be a very good choice in careers. Salaries are competitive, there is excellent job security, there are lots of interesting things to work on, and you can generally plan your career ladder to advance and sample a variety of positions.

However, a person does not have to be an employee of an agency to benefit from knowing something about government structures and operations. Jump forward many years and, after retiring early from the US Department of the Interior, I switched to teaching and conducting research at a university. When I arrived at the university I was astonished to learn that many faculty were not very knowledgeable about the operations of government agencies, even though they often received grants from these agencies. Some were not even sure what departments various agencies such as the US Forest Service came under (hint: US Department of Agriculture). Similarly, many faculty probably could not adequately describe the missions of most conservation-related non-government organizations (NGOs), despite perhaps working with biologists from these organizations. Communication, cooperation and interaction can be facilitated if you know how prospective partners think and what is important to them.

If you are considering a career in the conservation of natural resources, a course dealing with administration is very helpful. Major professional societies, such as The Wildlife Society, American Fisheries Society, or Society of American Foresters, have certifying programs designating men and women as professionals in their fields. Coursework for these certificates will include one or more courses in administration and policy. Some state agencies require that their new employees are certified, and others consider it a major plus. Also, certification can add to your prestige if you need to provide professional testimony in court or if you are seeking employment with an NGO.

Curriculum brings up another reason for this book. There are many books on environmental law, wildlife biology, forestry, fisheries, conservation and environmental science. In fact, there are entire departments that focus on these topics; perhaps you are in one of those now. While there are books on wildlife law, environmental law, environmental policy, and related topics, despite extensive searching I could not find any books that directly relate to the administration and policy of natural resources. To my knowledge, this current text occupies a niche that has not been previously filled.

IS NATURAL RESOURCE ADMINISTRATION IMPORTANT?

The goal of this book is to explain how government agencies and NGOs operate and what their various missions are. Before we discuss these commonalities, it might be good to define what we mean by natural resources and by administration. *Natural resources* are those things found in nature that can be used for the benefit of humanity. They include various sources of energy (fossil fuels, wind, solar, water), water, air, fish and wildlife, forests, land, minerals, and other factors that have value. *Value* is an entirely human concept; value can be determined by a monetary assessment or by having a function in the environment. It is hard to think of something in the environment that would have no value at all. Through the eons humans have become very ingenious, so the list of potential and realized resources is surprisingly long. Even rocks that were strewn through a field by glaciers (Figure I.1) or

FIGURE I.1 Stones scattered through a field may seem like a severe nuisance, but even these can be turned into a natural resource. *Credit: David Brown, Rocky ground, St Sunday Crag.*

volcanic eruption tens of thousands of years ago have value. Obviously, they could provide cover for a rodent from a soaring hawk. But even these may be used by humans. When the rocks are collected and stacked, they can form a very useful wall (Figure I.2A) or even a house (Figure I.2B).

Administration is the formal organization and operation of agencies, departments and other government and non-government bodies used by humans to coordinate and direct the conservation or protection of these resources. Administration differs from management in that management consists of activities directly associated with the conservation of a resource — management does, administration directs and organizes. In this book, we will focus on the administration, not the management, of natural resources.

Throughout the recent (since European colonization) history of North America, the concern for natural resources has steadily shifted from an early stage of dread to exploitation, to growing concern and preservation, to the

FIGURE I.2   Those rocks in Figure I.1 can be turned into a fence (A) or even a home (B).

application of scientific methods of conservation. Today, renewable resources in both the United States and Canada are generally in better condition than they were in the 19th and early 20th centuries (see Chapter 1. Briefly, beavers (*Castor canadensis*), white-tailed deer (*Odocoileus virginianus*), wild turkey (*Meleagris gallopavo*) and many furbearers and long-plumed birds are more numerous than they have been in the past. Many changes in laws and administration of natural resources accompanied and even preceded these shifts. Through laws, governments have directed management to enhance protection for these species and reintroduce them into formerly occupied areas. Bald eagles (*Haliaeetus leucocephalus*), brown pelicans (*Pelecanus occidentalis*), peregrine falcons (*Falco peregrinus*) and ospreys (*Pandion haliaetus*) have come back from near extinction after federal governments banned DDT. Other rare wildlife and plant species are receiving protection and management due to administrations passing the Endangered Species and Species at Risk Acts and the various Endangered Species Acts in provinces and states. In the United States millions of acres of National Forests have been set aside, and the boreal forests of Canada still rank among the top forests in the world due to administrative decisions made decades ago. The conditions that led to the Dust Bowl period in the late 1920s and 1930s are unlikely ever to occur again, as government agencies have established programs of education and soil protection on farmlands. Once-polluted rivers are again providing water for residential and industrial requirements.

All of this does not negate the need for continued conservation. While rare species are receiving protection, there are still nearly 400 vertebrates and over 1000 plant species that are endangered or threatened in the United States.[1] In Canada, the list of species of concern includes nearly 240 vertebrates and 165 plants.[2] In some areas of both Canada and the United States, over 90% of the wetlands present at the arrival of Europeans have been drained; overall, at least 50% of the wetlands in both countries are gone.[3,4] Through US history almost all old-growth forests, defined as a dominant tree class at least 150 years old, were hewed down decades ago, leaving less than 5% of the total in the United States. Although the amount of all forested land in the United States has remained virtually stable for the past century,[5] fragmentation is increasingly a problem, especially for forest-dependent birds.[6] The Society of American Foresters wrote a position paper in 2004[7] alerting the country that forested lands are declining in quality and are disappearing. In the past most of the converted forests went into agriculture but returned after agriculture declined or ceased in a region, thus the loss of forests was temporary. Increasingly, forested land is being permanently converted to urbanization, and Alig *et al.* 2003[8] have predicted that 23 million acres (9.3 million ha) may be permanently converted to pavement and developments by 2050. In addition to this, both nations are subject to uncertain threats of climate change.[9-11]

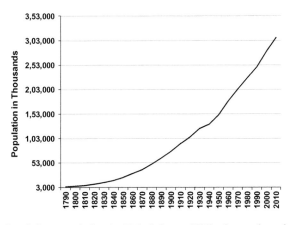

FIGURE I.3 Population growth for the United States. While the number of people in the United States continues to increase, the rate of increase is slowing; however, it will be many years before the population stabilizes. *Data Credit: US Census Bureau.*

A chief concern for the future of natural resources can be seen in Figures I.3 and I.4, which show population growth in the United States and Canada, respectively. The data from Canada are fragmentary in that the country did not have a national census until 1871, and regular 5-year censuses did not occur until 1956.[12] By 2015 and 2060, the estimated population in the United States will be 321 million and 420 million, respectively – at least a 31% increase.[13] Similarly, the estimated total population in Canada in 2015 will be 35 million, increasing to 52 million by 2060 – a 48% increase. Although Canada has around 11% to 12% of the total population of the United States, average population density, or the number of people per unit area, is not that different and is often a better measure than total population regarding the potential effects of humans on the landscape. More people per area implies greater urbanization, with greater loss of natural habitats and increased ecosystem stress. The population in the United States (Figure I.5) is dispersed compared to that of Canada (Figure I.6). While human density in the United States is highest along the eastern seacoast, California and the Great Lakes, in Canada the population is highly condensed in the southernmost 200 miles (~ 333 km). Midwestern and southwestern United States have less urbanization and more open land than coastal areas. For example, the northeastern United States has 11.2% of its land in urbanization, 11.6% in cropland, and 64% in other open habitat. The Midwest, exclusive of the Great Lakes, has 4.9% urbanization, 55.3% cropland and 30.8% in other open lands.[14] Whereas predictions for Canada's population suggest that the greatest growth through births and immigration will be in areas already heavily populated,[15] the population of the United States is steadily moving to the south and west.[13] As population densities shift, we can expect that areas now dominated

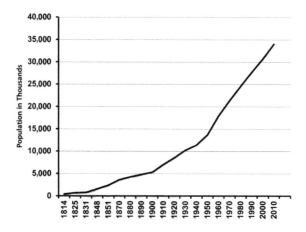

FIGURE I.4 Population growth for Canada. The growth pattern in Canada is similar to that of the United States, but note that Canada has far fewer people than the United States. *Data Credit: Statistics Canada.*

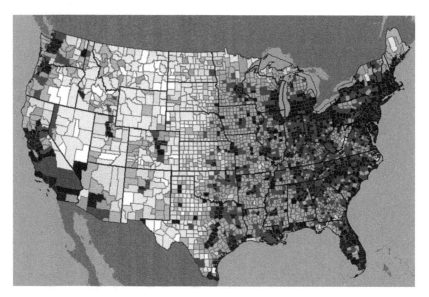

FIGURE I.5 Distribution of human population within the United States (the darker the color, the greater the population). While there are areas in the west that have low populations, coastal areas are crowded. *This figure is reproduced in color in the color plate section.*

by open land will gradually increase in urbanization. Northern Canada is dominated by boreal forest and tundra in part because the climate is too cold to grow farm crops. In addition to all of this, the demographic characteristics of humans in both countries are changing. Statistics show that the mean populations in both nations are getting older, living longer and having fewer babies.[16,17] What

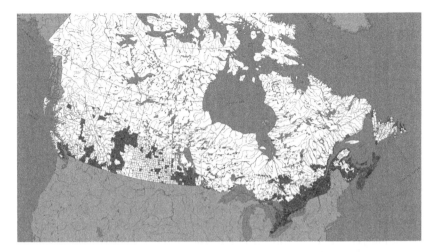

FIGURE I.6 Distribution of human population within Canada. Note that most of the people are clustered in the southern portion of the nation. *Credit: Statistics Canada. This figure is reproduced in color in the color plate section.*

this will mean to the demand for natural resources can only be guessed at this time. What we can predict is that the need for natural resources and the recreational opportunities they present will be different in the future than they are today. Natural resource agencies and NGOs will need to meet those challenges with new and innovative perspectives and ideas.

WHAT'S THE SCOPE OF THIS BOOK?

Our definition of natural resources encompasses all components of nature that have a value to humans. However, we are not going to cover all resources equally. We will focus on renewable resources. Emphasis will be on wildlife, fisheries, forests, parks, and their uses, including hunting, fishing, boating, forestry, nature observation, outdoor recreation and the like. We will spend less time discussing non-renewable resources and renewable sources of energy because their administration is becoming increasingly complicated, international, and regulatory in nature. Everyone recognizes that the price of gas at the pump is only partially affected by its actual value in the world market. Rather, politics, taxes and other factors are at least as important as real value in setting prices. Similarly, we will focus more on management agencies such as the US Fish and Wildlife Service than on regulatory agencies like the Environmental Protection Agency.

Some readers may wonder why I chose to include both Canada and the United States — isn't natural resource administration in just one of these countries sufficient for a book? Well, yes, natural resource administration in both countries is indeed very interesting and, written in great detail, each would provide adequate material for a book. However, Canada and the United States share a long common border; have similarities in their histories; share common attitudes towards conservation; are both federal democracies; have national and state or provincial levels that govern natural resources; and have an abundance of many types of natural resources. Thus, there are similarities in the way that natural resources are administered in Canada and the United States.

However, they differ from each other in ways that add interest to comparing them. The United States has a republican form of democracy with sharp separation of powers among the executive, judicial and legislative branches of government. This separation of powers extends to the structure of state governments. In the United States, the federal government takes a very involved role in natural resources and the lines between state and federal authority are sometimes blurred. Canada has a parliamentarian form of government, where the distinction between the executive and legislative branches is almost nearly invisible to those of us that live under a republican democracy. I have to admit that it seems strange that a legislator could also serve as the minister of an executive department and propose legislation that affects his or her own ministry. No doubt, the general election of a president who can be replaced every four years and has to share his authority with the legislature is a strange concept to many Canadians. In Canada almost all of the public lands are Crown lands, owned by the monarchy in England. According to Canada's constitution of 1982, rather stringently delineated authorities were given to the provinces and to the federal government on these Crown lands and the natural resources they contain. Regulatory authority for inland fisheries, wildlife, minerals and energy extraction, parks and forests within the provinces is virtually exclusive to the provinces. The federal government has jurisdiction over marine fisheries, all boat licensing, and the Crown lands and their resources in the northern territories. It seems that the Canadian federal government has been reluctant to cross the line between provinces and federal jurisdictions, even when natural resource issues are national in scope, but there is less hesitancy by the federal government in the United States. For example, the equivalent to the US Endangered Species Act (1973) is the Species at Risk Act, which was not adopted until 2002. The 30-year difference between enacting these critical laws appears to be, at least in part, due to Canadian federal resistance to usurping provincial rights. But wildlife and plants do not recognize state or provincial borders, and can be better protected through combined federal and state provisions. In contrast, public land in the United States is "owned" by the people – the citizens of the country.

STRUCTURE OF THE BOOK

This book is divided into four major parts, based on content.

Part I: Basics of Natural Resources

This section examines the fundamentals of natural resources, defines some of the basic terminology, and examines the history of conservation in the United States and Canada. In addition to describing what to expect in the book, this current chapter gets down to the real basics by defining what we mean by natural resources and administration as pertains to these resources.

In Chapter 1 we look at a variety of concepts such as conservation, preservation, sustainability, ecosystem services and adaptive management. A hundred and twenty years ago the primary concept of natural resources was preservation, or a "do not touch" attitude. From the landing of the first colonists to then much of our environment had been despoiled through overuse, abuse and contamination, and there was a growing concern for depleting resources. However, naturalists lacked knowledge about ecological principles or how to manage natural systems, so governments took about the only course open to them — which was to protect what there was. As knowledge increased, preservation gave way to conservation. Scientists learned that natural resources could be used wisely without destroying their integrity or diminishing their base. Of course, non-renewable resources such as oil and natural gas cannot be replaced once they are used, so conservation for these resources meant decreasing waste and enhancing efficiency of extraction.

Some 40 years or so ago the buzz word for conservation became *biodiversity*. This concept stressed that greater biodiversity led to greater ecosystem stability, but the associated concern was and continues to be that the number of species is diminishing rapidly. Because human activities such as deforestation, extensive plowing and widespread contamination have caused many of these declines, some look upon humans as an enemy to natural systems rather than the solution to environmental problems.[18] While the importance of biodiversity as a measure of environmental health has not diminished, conservation ideas are changing to an understanding that humans are here for the long run so let us make the best of it. Concepts such as *sustainability* and *ecosystem values* have been added to biodiversity as buzzwords. If humans are going to be here, then let us do what we can to minimize the environmental damage caused by their activities — in fact, let us develop programs of learning so that humanity will develop a greater sense of stewardship and acceptance of its role in the environment. So we have sustainable agriculture, sustainable forestry, sustainable energy production and use, and so on. This approach does not replace biodiversity as a measure of environmental health, but is arguably more realistic than banking entirely on that premise.

Chapter 2 takes a close look at the history of conservation in the United States and Canada. We focus primarily on wildlife conservation because its history with humankind reflects that of natural resources in general, and there is considerable information available on this topic. Much of the history of Canadian natural resource legislation has paralleled that of the United

States, with a lag period of about 20−30 years. Politics may be involved with this lag, but we should also acknowledge that Canada's smaller population may have slowed the rate of anthropogenic environmental damage which, in turn, facilitated the lag between the nations. For the most part, Canada and the United States have made great progress in the management and concern for natural resources. We certainly don't have all the answers, but we have come to know a great deal since the English, Spanish and French colonists first arrived.

Part II: Environmental Law

Chapter 3 demonstrates that in many ways the different forms of government in Canada and the United States can be observed in the historical progression of environmental laws, but it is very interesting that both nations have converged in societal concepts about natural resources. In the United States, the early history of environmental laws gave precedence to the states. The doctrine of Public Trust with regard to wildlife in the broad sense was formulated with one of the earliest Supreme Court decisions, and exists until this day. The doctrine can be succinctly expressed as "no one owns a wild animal until it is reduced to possession". It is when the animal is in possession of a human that our laws can determine if taking that animal was legal. Another way of expressing this is that wildlife is "owned by everyone and no one in particular". The concept of Public Trust pertains in both countries, and through time has led to the North American Model of Wildlife Conservation. While wildlife is owned by society in general, the laws of ownership of other natural resources are very different. For example, minerals, fossil fuels, and lands such as forest, grasslands or cultivated lands, can be privately owned. Thus, the trees in a forest, arable land in a grassland, or a pond enclosed by private lands can be owned outright. Any fish in that pond also are the property of the landowner. However, if the same species of fish is in a body of water that flows through a region or in a lake surrounded by multiple property owners, the fish are treated by the Public Trust Doctrine in the same way as wildlife. Depending on the state, mineral rights either come with the purchase of property or can be owned separately. On public lands, including federal, state and Crown lands, the concept is that resources are administrated for the good of all, although this is not always done with perfection.

Chapter 4 is a more intense examination of specific federal laws dealing with natural resources. Individual states and provinces have their own laws, and it would take a much larger volume to report on all the particulars. At this point, it may be worthwhile to remind ourselves that these laws have been enacted to protect us from ourselves. It seems to be human nature to take more than one ought, for federal, state and provincial laws exist to assure that there are sufficient resources for all.

Part III: The Bureaucracy of Natural Resources

During the development of conservation, the legislatures in the United States and in the House of Commons in Canada have established agencies and departments to administer these natural resources. At the same time, states and provinces have established their own agencies that in many ways mirror federal counterparts. Descriptions of these agencies, including specific details of their histories and authorities, their missions, goals and business operations, budgets, and infrastructure are presented in this chapter. The focus is on the federal level because, once again, a detailed inventory of state and provincial agencies would be excessive.

Chapter 5 includes accounts of Canadian ministries that oversee natural resources. Compared to the United States, Canadian federal bureaucracy is more condensed and often more involved in support functions to the provinces than in actual management of resources. Again, these approaches are due to constitutional differences in the distribution of federal and state (provincial) authorities between nations. There are three principal ministries that deal with conservation issues. Environment Canada has sub-ministries dealing with weather-forecasting for the nation, waste management, pollution, wildlife, sustainable ecosystems, water resources, biodiversity and enforcement. It also houses the national park system as an independent agency supervised by the same minister of the environment. Natural Resources Canada is the principal Canadian federal agency for energy, forests, minerals, metals and earth sciences, topographical mapping, and remote sensing. Fisheries and Oceans oversees marine fishing, commercial freshwater fisheries and regulates shipping and boating.

The US department most involved with fisheries, wildlife, parks, minerals and mining, and grazing lands is the Department of the Interior, which is covered in Chapter 6. The US Fish and Wildlife Service is the lead for endangered species conservation in the nation; it also encompasses the national wildlife refuges, most of the federal conservation law enforcement authority, and ecological services offices across the country that provide guidance on a host of conservation topics, including contaminants, wetlands, endangered species and legal matters. The National Park Service manages the National Parks in the United States, but also oversees the various monuments, national historical sites and national parkways. The Bureau of Land Management has oversight responsibilities for millions of acres of grazing lands in the west. The US Geological Survey, which for a long time was composed of sections dealing with minerals, monitoring of waterways including major rivers, and cartography, now in addition conducts all of the biological research activities for the Department. The USGS also houses the Cooperative Research Units that provide educational opportunities for graduate students across the country. The other agencies within the Department deal with Amerindian affairs, and mining regulations on land and in the sea.

The US Department of Agriculture (Chapter 7) has oversight responsibilities for various farm support programs. However, it also contains two agencies critical to natural resource conservation in the country. Among other activities, the US Forest Service administers millions of acres of national forests and grasslands, and provides support to states and private concerns on forest management. The National Resource Conservation Service provides guidance on prevention of soil degradation and erosion to producers, and professional guidance to a variety of farmland-associated programs, including the Conservation Reserve Program, Environmental Quality Incentive Program, Swampbuster, Sodbuster, and others. The agency works with the Farm Services Administration in the financing of these programs.

Chapter 8 deals with a very diverse topic: how natural resources are handled by the various states and provinces. Any obvious differences between the United States and Canada in this regard are pretty much concealed by the greater differences that exist *among* states and provinces. Some states, for instance, have separate departments for each natural resource, while others have unified departments of natural resources with separate sub-agencies dealing with fish and wildlife, minerals, state parks, etc. Similarly, each province has developed its own system for managing natural resources. We also discuss the ways in which states obtain funding for their conservation operations.

Part IV: Non-Governmental Agencies, People, and Money

Chapter 9 examines an even more diverse array of conservation organizations — non-governmental organizations, or NGOs. Non-governmental organizations are private groups that support specific conservation objectives or purposes. In many ways they connect government agencies and their policies with the general public. NGOs receive their funding from membership dues, investments, sale of books, clothing, and other items, and donations. As the chapter shows, you can probably find an NGO for just about any topic you can consider. In the conservation arena there are huge, multi-million member NGOs all the way down to small organizations consisting of a few dozen members. Most NGOs have their own areas of specialty and can be separated into different mission categories. For example, some are societies that bring together professional scientists, biologists, etc., in wildlife, fisheries, forestry, parks or other occupations. Others focus on a particular species, such as butterflies; many in this category are sportsmen's organizations that promote a harvestable species such as white-tailed deer, elk (*Cervus canadensis*), trout, or ring-necked pheasant (*Phasianus colchicus*). Other categories include diverse organizations interested in national and international conservation, sustainability, and the poor. Among the other types of NGOs are those that act as shills, having environmentally friendly sounding names but actually supporting eco-damaging businesses, and radical groups that believe that the ends do justify the means, regardless of what those means may be.

One common denominator to all forms of natural resource management is people. Field-level natural resource managers work with people directly on an everyday basis. We can label these people as stakeholders, users, customers, clients, or just "the public". They may be hunters at a refuge, bird watchers at a sanctuary, campers, hikers, ATV users, or other types of clientele. The public can be great supporters of conservation through donations, writing letters to their elected officials, volunteering in clean-up campaigns, or staffing non-professional slots. They can also be a major source of problems, such as dumping garbage, causing trail erosion, overfishing or overhunting areas, getting lost or injured, violating regulations, etc., etc. Nevertheless, agencies, whether private or governmental, exist because of these people. It has been said natural resource management is 90% people management and 10% resource management. Therefore it pays for agencies to know the public they are likely to encounter, to strive to make this public informed about what the agency does, and to maintain good relations with the public. Three of the chapters in Part IV of the book examine aspects of dealing with the public, while the fourth chapter looks into a crystal ball to see what the future of natural resource conservation may be like.

Chapter 10 discusses the past and current attitudes of Americans and Canadians towards conservation. A lot of examples focus on wildlife and fish, for that is where most of the studies have occurred. What trends are apparent in fishing and hunting? What stimulates non-consummatory users of nature? How do social factors such as gender, ethnicity, age, or socioeconomic status affect attitudes towards conservation?

Chapter 11 presents information on how agencies work with the public through public relations departments. Case histories are presented to illustrate the importance of knowing your public, and how to present effective programs to shape public attitudes through education and building understanding between local communities and the adjacent conservation site.

Chapter 12 deals with that all important commodity, money. Given that there is never enough money to do everything an agency would like to do, it is important to know something about the budgetary process and alternative sources of funding. In recent years planning has become mandated at all levels and we spend some time talking about this pervasive activity.

Finally, Chapter 13 provides a brief review of the book and discusses future issues, including climate change, the high extinction rate, sustainability, and human equality.

REFERENCES

1. US Fish and Wildlife Service Endangered Species. <http://www.fws.gov/endangered/>.
2. Committee on the status of endangered species of wildlife in Canada. <http://www.cosewic. gc.ca/eng/sct0/index_e.cfm>.

3. Dahl TE. *Status and trends of wetlands in the conterminous United States, 2004–2009*. Washington DC: US Fish and Wildlife Service; 2010:http://www.fws.gov/wetlands/Status-And-Trends-2009/index.html

4. Badiou PHJ. Conserve first, restore later: a summary of wetland losses in the Canadian prairies and implications for water quality. <http://Yorkton.ca/news/2013/waterseminar/pdf>; 2013.

5. Alvarez M. *The state of American forests*. Bethesda, MD: Society of American Foresters; 2007.

6. Schnell JK, Harris GM, Pimm SL, Russell GJ. Quantitative analysis of forest fragmentation in the Atlantic forest reveals more threatened bird species than the current Red List. *PLoS ONE*. 2013;8(5):e65357.

7. Society of American Foresters. Loss of United States forest land: A position statement of the Society of American Foresters, Bethesda, MD. Available at <http://www.forestry.org/media/docs/ak/SAF_National_PositionStatement-loss_of_forest_land-exp-20141205.pdf>; 2004.

8. Alig RJ, Plantinga AJ, Ahn S, Kline JD. Land use changes involving forestry in the United States: 1952 to 1997, with projections to 2050. Gen. Tech. Rep. PNW-GTR-587, Portland, OR: USDA Forest Service, Pacific Northwest Research Station.2003:92. Available online at: <www.fs.fed.us/pnw/pubs/gtr587.pdf>.

9. Andeson-Teixeira KJ, Miller AD, Mohan JE, et al. Altered dynamics of forest recovery under a changing climate. *Global Change Biol*. 2013;19:2001–2021.

10. Jenouvrier S. Impacts of climate change on avian populations. *Global Change Biol*. 2013;19:2036–2057.

11. Frelich LE, Reich PB. Will environmental changes reinforce the impact of global warming on the prairie-forest border of central North America? *Frontiers Ecol Environ*. 2013;8:371–378.

12. Statistics Canada Estimated Population of Canada. <http://www.statcan.gc.ca>.

13. US Census Bureau Population Projections. <http://www.census.gov/population/projections/data/national/2012/summarytables.html>.

14. Nickerson C, Ebel R, Borchers A, Carriago F. *Major uses of land in the United States, 2007*. Washington, DC: EIB 89, USDA Economics Research Service; 2011.

15. Beaujot R. Projecting the future of Canada's population: Assumptions, implications and policy. *Can Stud Pop*. 2003;30:1–28.

16. Statistics Canada Demographic Change. <http://www.statcan.gc.ca/pub/82-229-x/2009001/demo/int1-eng.htm>.

17. US Census Bureau. National Population Projections: Summary Tables. <http://www.census.gov/population/projections/data/national/2012/summarytables.html>; 2012.

18. Foundation for Deep Ecology. The Deep Ecology Platform. <http://www.deepecology.org/platform.htm>; 2012.

Part I

Basics of Natural Resources

Chapter 1

Differing Perspectives on Natural Resource Policy

INTRODUCTION

Suppose we develop a scenario where you are the head of a natural resource agency. A wealthy business person has donated a 5000-acre (~ 2025 ha) tract of land to your agency. After taking an inventory of the property, you have a few questions that need to be addressed. What are you going to do with this property? Does it contain a rare, fragile type of habitat or a species of concern to make you want to preserve the land? If you decide to manage it, what are your conservation goals going to be? Will you try to maximize the number of species on it, try to enhance its sustainability, increase its resilience, or do you have other goals for it? Once you have carried out your

D.W. Sparling: Natural Resources Administration. DOI: http://dx.doi.org/10.1016/B978-0-12-404647-4.00001-5

conservation plan, how will you evaluate if the plan was effective? Will you just assume that it was effective, or will you establish some mileposts and criteria to determine objectively if the plan was successful? All of these are real issues that natural resource managers and their administrations encounter on a regular basis. This chapter examines some of the ways in which organizations approach these natural resource management and administrative issues.

PRESERVATION AND CONSERVATION

These two terms are often used interchangeably but really have very different meanings. *Preservation* can be defined as the protection of resources, whether they are land, species, specific genotypes, landscapes or some other factor. Preservation involves little to no direct management — not quite an entirely "hands off" approach, but close. If preservation is associated with land, access to the area is often restricted to reduce human disturbance. If it is a rare species, access to its habitat may be similarly restricted or an agency may decide to reduce disturbance as much as possible, or the species may be so rare that it will either have to have special protection or be relocated entirely to captivity until enough information is obtained about its ecological needs that it can be restored to the wild.

Père David's Deer (*Elaphurus davidianus*), also known as the *milu* (Figure 1.1), for example, is native to China but became extinct in the wild over 1000 years ago.[1] For centuries a small population was maintained in the Nanyuang Royal Hunting Garden in Nan Haizi, near Beijing. A small group of these deer had been smuggled out of China and brought to Europe during

FIGURE 1.1   Père David's Deer (*Elaphurus davidianus*) at Sharkarosa Wildlife Ranch in Pilot Point, Texas — a species that could serve as the poster child for protectionism.

the 1800s. During the Boxer Rebellion of the late 19th century the deer at Nanyuang were slaughtered for meat, and the species would have become extinct had those animals not been transported to Europe. In 1985 a few animals from the European group were used to redevelop a Chinese population at Beijing Milu Park, the site of the old Nanyuang Gardens. A few years later a second population was established at Dafeng Nature Reserve.[2] In 1993, and again in 2002, some deer from both parks were released into Nature Preserves in China. After more than 1000 years, the species was back in the wild, although still heavily protected. During all this time the species was being *preserved*, because very little active management was occurring. Arguably, moving some deer into the Nature Preserves might be interpreted as conservation, but the activity was minimal.

A second example of a more local species that requires special preservation is the California Condor (*Gymnogyps californianus*). This bird is a large scavenger of southern California, with a body weight of 22 pounds (10 kilograms) and wing span of up to 9.5 feet (2.9 meters). The species is ancient, with fossil evidence going back at least 100,000 years to the mid-Pleistocene epoch. At one time California condors occupied the western seaboard states, Baja, British Columbia, Texas and New Mexico, and had populations in Florida and the Northeast.[3,4] The species declined precipitously with the extinction of large mammals during the late Pleistocene.

In recent times condors were shot and poisoned as vermin, and fell victim to secondary toxicity from lead bullets in deer, wild pigs, and antelope left by hunters. In addition, museum officials collected eggs from nests so as to have remnants of a vanishing species. Lead shot poisoning was the principal culprit in recent declines of the species.[5] Condors also have low fecundity, which does not help their recovery. They mature at five to seven years, lay only a single egg, and typically breed every other year, although sometimes they may breed in consecutive years. Adults in the wild may live 40 years or more, and the oldest one in captivity was 71 years old when it died.[6] In 1982 only 22 birds existed in the wild, and the US Fish and Wildlife Service decided to bring all surviving birds into captivity until they could determine what to do with the species. By 1987, there were no condors left in the wild.[3] This is the preservation portion of the actions taken to save the species. However, the story also involves conservation, in that many of the Condor pairs reproduced and biologists would hand-rear hatchlings (Figure 1.2). After eight years in captivity the population had grown sufficiently to allow limited releases. Over the years, condors have been released and monitored in the Grand Canyon area of Arizona and other parts of the Southwest. These released birds are considered experimental and therefore expendable. Other management activities included training condors that are to be released to avoid electrical lines, and passing a law that forbids the use of lead ammunition in condor release sites. Today there are about 300 condors in existence, with around 70 free-flying in the West.[7]

FIGURE 1.2   A California Condor chick (*Gymnogyps californianus*) being tended by a human wearing a glove that looks like the head of an adult condor.

Many non-governmental organizations (NGOs) of various sizes strive to protect and preserve unusual habitats from development. Small NGOs include Green Earth Inc., in southern Illinois,[8] which owns six small tracts of woodland and bottomlands totaling 220 acres (89 ha) in a semi-urban environment. These small preservationist groups often have a knack for finding isolated gems of nature in otherwise developed areas, and protecting them from destruction (Figure 1.3). Much larger organizations can protect natural areas of international importance. For example, the Nature Conservancy and the World Wildlife Fund have combined their resources to protect, among other areas, a 150,000-acre (60,702 ha) Valdivian Coastal Reserve in Chile that houses ancient coastal forests and a multitude of wildlife[9] (Figure 1.4). In the United States and Canada national parks, monuments and wilderness areas practice the concept of preservation while still allowing people to visit these areas, even though visitor activities may strongly influence the pristine quality of these parks.[10,11]

A problem with preservation of an ecosystem or landscape context, however, is that natural areas are not stagnant. Succession may advance the natural development of the area, and events that set succession back, such as fires, can change the conditions of the area being protected. If natural changes are acceptable by human decisions, then preservation can continue. On the other hand, if it is more desirable to maintain an area in a particular set of conditions, conservation with active management will almost always be necessary.[12]

FIGURE 1.3   Chautauqua Bottoms, a preserve owned by Green Earth Inc., one of hundreds of small non-government organizations created to protect ecologically unique sites. *Reproduced with permission of Green Earth Inc., Carbondale, IL.*

FIGURE 1.4   The Valdivian Coastal Reserve, Chile. This is a 150,000-acre (60,700 ha) preserve owned by the World Wildlife Fund and Nature Conservancy. *Credit: Nature Conservancy, http://www.nature.org/ourinitiatives/regions/southamerica/chile/index.htm.*

A third and very different example of modern-day preservation can be found in the many gene banks that have arisen around the world. Some of these banks are huge, temperature-controlled buildings that house millions of seeds and other tissues. The National Center for Genetic Resources Preservation, under the US Department of Agriculture, for example, maintains stocks of plant seeds, germplasm, and cryogenically preserved embryos of animals and plants (Figure 1.5). The stated mission of the Center is "to acquire, evaluate, preserve, and provide a national collection of genetic resources to secure the biological diversity that underpins a sustainable US agricultural economy through diligent stewardship,

FIGURE 1.5 A panorama of seeds stored by the USDA's National Center for Genetic Research and Preservation. *Reproduced courtesy of the USDA National Center for Genetic Research and Preservation.*

research, and communication".[13] Some of these repositories have been established so that if a cataclysm (whether of natural or human origin) occurs there will be organisms that can be used to get a new start − provided that there are human survivors, of course. Other repositories, such as the USDA Center, exist because scientists and the public see an intrinsic value in preserving potentially valuable genotypes that now populate this planet.

While these are a couple of ways that we are using preservation today, the concept was very important during the late 1800s and early 1900s. At that time Americans and Canadians began to be aware that many of the natural resources, such as forests and wildlife, that we had taken for granted were rapidly disappearing (see Chapter 2). The American bison (*Bison bison*), beaver (*Castor canadensis*), passenger pigeon (*Ectopistes migratorius*) and many other once common species had become very rare. Old growth forests were gone in the eastern United States, and greatly reduced in the West. Urbanization was overtaking near-pristine habitat at a rapid pace. Following the concept of Manifest Destiny, people were moving out West in large numbers. In the late 1800s, and even in the first third of the 1900s, we really did not have the scientific knowledge on how to manage these dwindling resources, especially wildlife and fish. Therefore, we did what we could, which was to protect those resources that humans believed to be most ecologically or economically valuable. During this period, both countries set aside some very important habitat as wildlife refuges, national, state or provincial parks, and forests.

In contrast to preservation, *conservation* implies active management of natural resources and has replaced strict preservation as the more common natural resources activity. The classical definition of conservation, "the wise use of natural resources", has been ascribed to Gifford Pinchot.[14] Conservation has also been defined as the prudent use of natural resources with an eye to the maintenance of future availability and productivity.[15] By its definition, conservation explicitly states that natural resources should be used, but with an understanding and concern for the perpetuation of the resources. Wanton waste through such activities as burning and slashing of forests, overharvesting of wildlife, pollution, abuse of soil, and unmitigated extraction and burning of fossil fuels are not part of conservation. Today, more so than ever, conservation involves recycling renewable resources and finding ways of reducing the exploitation of non-renewable resources. We have gained a wealth of information on how to manage species, habitats and resources; all of this knowledge must be applied to assure that we have sufficient and varied natural resources now and in the future.

GOALS OF CONSERVATION

Conservation has become an everyday word and practice. We constantly hear about how we should conserve this and recycle that, and use energy efficient devices (Figure 1.6) and hybrid automobiles. These are great advancements and they do make a difference. However, we will use "conservation" in a more scientific context.

"Your bright ideas don't seem as bright as they used to."

FIGURE 1.6   Cartoon illustrating the future of energy conservation. *Credit: Randy Glasbergen* *http://www.glasbergen.com.*

Given that conservation is more widely used than preservation in modern natural resource management, what are the goals that agencies and, by extension, the public that they serve typically direct their conservation efforts towards? What are the target points for today's conservation? Michael Soulé[16] was among the first to address this question, and described *conservation biology* as a "new stage in the application of science to conservation problems" (Soulé, 1985, p. 727)[16]. He called conservation biology a "crisis" science because at that time conservationists often had to react to damage that was already done or to harmful activities while they were occurring, and they had to act with only a partial understanding of the effects these activities might exert. Conservation was more reactionary than preventative. Now, as then, uncertainty has to be tolerated in conservation, but we have come some distance in reducing this uncertainty. Soulé listed diversity of organisms, ecological complexity, and biotic diversity as the goals of conservation.[16] Today, we might lump all three elements under the single title of *biodiversity*. Subsequently, Soulé wrote on factors that tend to diminish biodiversity, including habitat fragmentation under urban[17] and natural[18] conditions, the keystone species concept,[19] and other issues dealing with conservation biology.

While biodiversity remains a leading goal for conservation, through the years other goals have been stipulated; they can involve individual species,[20] whole groups of species,[21] or human/species interactions.[22] Specific habitats,[23] habitat quality[24] and a plethora of other topics are chief concerns. As we will see, many conservationists are also emphasizing human economic goals in their conservation priorities.

Multiple Use, Sustained Yield

Federal agencies both in Canada and the United States have specific missions that are conducive to particular types of conservation. For example, the US Forest Service, Bureau of Land Management, many state and provincial Department of Natural Resources or their equivalents, elements of Environment Canada, and Natural Resources Canada operate under a *multiple use, sustained yield* form of conservation. This concept was first fostered in the United States by Gifford Pinchot, and further information on the philosophy can be found in various chapters throughout this book (see, for example, Chapter 7). *Multiple use* is the idea that the public should be able to use natural resources in many different ways. For example, national, state and provincial forests are open to many different interest groups, including hikers, campers, hunters, fishermen, kayakers, and other sportsmen. However, many are also open to all terrain vehicles (ATV), snowmobiles and even "harder" uses of natural resources, such as timber cutting, and sometimes mining and oil exploration. It can be a great challenge to make sure that all of these uses can occur in one area without interfering with each other. Hikers, for instance, most likely find ATVs bothersome, and camping may not be compatible with timber harvesting.

Keough and Blahna cited eight characteristics that go into management for multiple use or, as they called it, integrative collaborative ecosystem management: (1) integrated and balanced goals; (2) inclusive public involvement; (3) stakeholder influence; (4) a consensus group approach; (5) collaborative stewardship; (6) monitoring and adaptive management; (7) multidisciplinary data; and (8) economic incentives.[25] Note that all of these elements deal with stakeholders or economics in some way, which supports the contention that natural resource management is people management. The astute manager/administrator will solicit public opinion and work towards a consensus from the initial planning stages of any multiple use project. The more uses, the broader the stakeholder base that has to be brought together.

The second part of this philosophy is *sustained yield*. This is a mandate that, to the best of a manager's ability, resources should not be depleted or abused − that they should last into perpetuity. This may be more easily accomplished with hiking trails than with surface mining, but effective conservation needs to maintain a supply of renewable resources through setting limits and quotas to their harvest and implementing active management to replace what is harvested. Non-renewable resources, by their very definition, will be depleted through extraction, so the goal here is to prolong their availability. In all cases, waste and spoilage should be avoided.

Use-Constrained Management

In contrast to the multiple use, sustained yield form of conservation, many agencies insist on more limited use of their lands and the resources they produce. For example, national wildlife refuges and parks in the United States and Canada restrict the use of their properties to low-invasive purposes. While most allow hunting, hiking, and fishing, many disallow ATV and snowmobile use, greatly curtail timber cutting except for habitat management purposes, and prohibit oil, gas and mineral exploration or extraction. Still others, such as the national parks in Parks Canada and the National Park Service in the United States, promote recreational use but prohibit or limit harvest of fish, plants or wildlife, and attempt to maintain their lands in as pristine a condition as possible; thus, their goal is to operate closely to a preservation model. The chief goals of these areas include maintenance of a natural area *and* visitor enjoyment. The two goals are not always mutually compatible. For example, in at least one study there seemed to be more wildlife road-kills in protected areas than in non-protected areas, with amphibians and reptiles particularly hard hit in protected areas.[26] The authors suggested that protected areas actually receive more visitations and greater road traffic per mile of road than many non-protected areas. Plants are also frequently trampled along trails.[27] Roads used for visitor access may also increase the risk of invasive plant species entering a protected area.[28] Interestingly, visitors were often quite resistant, even hostile, to park restoration efforts, for the temporary human disturbance decreased the value of their experiences.[29] It is obvious that visitor education plays a very important role in keeping natural areas natural.[30]

Human-Focused Conservation

Conservation Biology or Conservation Science?

Soulé (1985) distinguished conservation biology from natural resource fields of fisheries biology, forestry and wildlife management in that conservation biology focuses on the environment whereas the other three areas are more utilitarian and include economics.[16] Scientific journals such as *Conservation Biology*, *Journal of Nature Conservation* and *Biological Conservation* generally publish articles with a focus on conservation for the sake of the ecosystem or landscape. Conversely, *Journal of Wildlife Management*, *Transactions of American Fisheries Society* and *Journal of Forestry* publish articles which are more pragmatic in that they deal with resources used by humans. So there would seem to be a division within the broad field of conservation. However, there are journals, such as *Biodiversity and Conservation*, *Global Climate Change* and *Environmental Conservation*, that publish papers on several kinds of conservation issues. Note that none of the above journals is exclusively of one type or another, but there is a tendency for journals to specialize in their areas of conservation. This diversity supports Kareiva and Marvier's[31] attempt to clarify some of this current apparent division by coining the term "conservation science" (as opposed to Soulé's *conservation biology*). According to Kareiva and Marvier, conservation science is an advancement of Soulé's conservation biology in that it is based on better information about the systems we are conserving, has both ecological and social dimensions, and has as a key goal the improvement of human well-being through the management of the environment. Conservation science recognizes that ecological dynamics can no longer be separated entirely from human dynamics. Every conservation action requires human input in the form of decisions and economics, and most have outputs that directly affect humans.

Early conservationists did not have to be as concerned about public opinion as they are today. In 1872, when conservationists established the first national park with Yellowstone, they did not consult the American public; rather, a very small body of interested people went directly to Congress and persuaded legislators to create the park.[32] Similarly, Banff was designated a national park in Canada in 1885 without widespread public input. Today, the US National Environmental Policy Act (NEPA), Canadian Environmental Protection Act (CEPA), and mass media prevent governments in Canada and the United States from acting unilaterally on environmental issues. This is good for the people who pay for government action and should have a voice in how their money is spent. Conservation science also utilizes the wealth of information that we have obtained through experience in making wiser, more efficacious decisions.

Everyone can be players in this scenario, even corporations. Kareiva and Marvier[31] stated that rather than being major deterrents to a healthy environment, corporations can become more actively involved as positive forces through funding and responsible utilization of resources. The authors also explained that conservation practices should not interfere with human rights. It is not ethical to force people off their lands for the sake of conservation, especially without adequate compensation and alternatives for them to continue their way of life. Many times the people who are unfairly affected are the poor and disenfranchised, with no say in the decision-making process. As the authors concluded, "People deserve a voice in their own fates as well as in the fates of the lands and waters they rely on. Not only is this arguably the right thing to do from an ethical perspective, it will probably improve conservation outcomes" (Kareiva and Marvier 2012, p. 967)[31] In this regard, several conservation NGOs have published statements on human rights and formed a Conservation Initiative on Human Rights.[33] Ideas such as these place us in a good position to understand two other concepts of present day conservation: sustainability, and ecosystem services.

Sustainability

There is no universally accepted definition for the modern, conservation-based definition of *sustainability*. While related to the concept of sustained yield fostered by Pinchot, sustainability has evolved beyond that point and has been applied to a wide array of human needs and activities. Pinchot was primarily concerned with developing a system of silviculture that would allow a forest to be harvested over a prolonged period of time. Similarly, sustained yield has been applied to other renewable resources, such as in fisheries and wildlife management. Today, however, the concept of sustainability involves the resource and its value to human society. In 2000 the United Nations produced a *Millennium Declaration*,[34] which outlined the present and future needs of human development. Among these needs were an end to poverty, support for human rights, and a section on the environment that expounded prudent management of all living species and natural resources, in accordance with the precepts of *sustainable development*. A 2005 *World Summit on Social Development*[35] that reviewed the progress of the Millennium Declaration noted that sustainability requires the reconciliation of environmental, social equity and economic demands — which were defined as the "three pillars" of sustainability. So, ecological or environmental sustainability includes an international desire to provide for the common human good in terms of the environment.

One of the many ways to represent the concept of sustainability is by three concentric circles of environment, society and economy.[36] The environment is all encompassing, and includes humans. In turn, human society includes economics, so each of the three factors is seen as being dependent on the next higher level (Figure 1.7). Another, somewhat more complex, explanation[37] shows the three factors as mutually overlapping (Figure 1.8). When two factors intersect we see situations that gradually improve from bearable to equitable to viable, but it is only when all three intersect that sustainability occurs, as indicated by the **S** in the diagram.

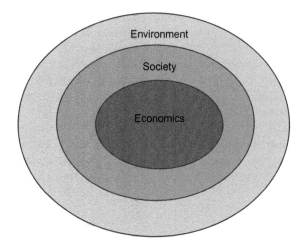

FIGURE 1.7 A three concentric circle design to explain ecosystem sustainability. The smallest circle, economics, is part of and dependent on the larger element of human society, which in turn is dependent on the totality of the environment.

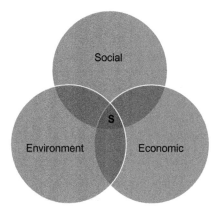

FIGURE 1.8 An alternative explanation of sustainability, where economic, social and environmental factors intersect. Each has effects on the other two elements but sustainability occurs where the three converge.

Becker[38] used three other terms to explain sustainability. The first of these is *continuance*—which simply means to keep going, to maintain. A sustainable activity is one that can be perpetuated. For example, setting quotas on the harvest of ocean fishes should be done so that fish stocks are maintained. A sustainable ecosystem is one that can maintain its chief characteristics over an indefinite time period. Society has to agree that these fish stocks or ecosystem characteristics are desirable, and that they are willing to invest time and money to maintain these characteristics. Also associated with sustainability is that fact that society is already *oriented* towards stability and some degree of permanence. Sustainability is already considered desirable by society, and is regarded as a major objective in human activities. A third aspect of sustainability is that it has an element of *relationship*. Sustainable programs aim to satisfy the natural resource needs of the present generation while not hindering the ability of the next generation to also meet its own needs. Thus, there are relationships within the current generation of society and between the current and future generations.

The concept of sustainability has been applied to just about all aspects of the environment as it intersects human life. A recent search on "sustainability" and "environmental sciences" with a popular scientific search engine,[39] for example, retrieved nearly 12,800 articles, the earliest going back to 1981. These articles deal with marine resources, wildlife, fisheries, forestry, water, agriculture, tourism, carbon foot prints, solar power and other renewable resources. Non-renewable resources such as oil and gas extraction were also well represented. Some of the more practical articles wrote on sustainability in human neighborhoods,[40] low income housing,[41] and even home waste disposal units.[42]

ECOSYSTEM SERVICES

Remember the 5000 acres of land that your "agency" received at the beginning of this chapter? Suppose there was a pond on that property. Not a huge pond — maybe about an acre. How would you respond if someone asked, "Why is that pond desirable, why not just fill it in and plant rutabagas on it?" You might respond that you don't like rutabagas, but a more refined answer might address some of the practical aspects of the pond. During wet seasons the pond retains water that would otherwise run off into a nearby river and add to flooding. The pond, if it is deep enough, may contain fish which can be used recreationally or as a food source. The willows in and around it produce oxygen and store carbon dioxide that would otherwise contribute to greenhouse gasses. You might also address aesthetic values. Many people like to listen to the frogs at night, or enjoy watching the herons and blackbirds the pond attracts. All of these are characteristics of the pond that in one way or another — flood prevention, food, carbon sequestration or aesthetics — have value. You might argue that placing an economic value on watching wading birds is a bit ridiculous, but studies show that home values are enhanced by the presence of green space and bird diversity[43] and more than 11 million people watch birds each years in the United States, not including hunters and anglers,[44] so bird watching has considerable value. In similar ways, a forest provides many benefits, including but not limited to carbon sequestration by the trees, water retention and purification through porous soils, temperature amelioration both in and around the forest,[45] timber, and wildlife habitat. Everyone knows that honeybees (*Apis mellifera*) produce honey, but the value they and other pollinators provide to all sorts of plants and crops is far more valuable, amounting to $14 billion annually just in the United States.[46]

These assets associated with a pond, a woodlot, bees or other resources are called *ecosystem services*, and one goal of conservation is to maximize or at least maintain the value of these ecosystem services. We live in a world with a very rapidly growing human population. In the 55 years between 1950 and 2005, the world's population increased by 4 billion people. In 2011 we had just around 7 billion, and the United Nations predicts that the population could be 10.9 billion in 2100 before it levels off to around 9.2 billion.[47] Most of this growth will occur in developing countries, where the greatest biodiversity occurs. As population growth continues, there will be greater demand for resources of all types. Back in 1997, the global value of ecosystem services was estimated at around $33 trillion per year.[48] Such total estimates are interesting, but have many underlying assumptions and variables. deGroot *et al.*[49] provided a more detailed analysis for particular biomes, and presented value ranges based on the quality of the biome. The authors did not calculate a global total, but found that the value of services provided by open oceans had a median value of 490 int$/ha per year whereas coral reefs had a median value of 350,000 int$/ha per year; the highest quality coral reefs could yield over 2.1 million int$/ha per year in ecosystem services. The value of the int$ (international dollar) the authors used was the purchasing power of the 2007 US dollar which today would yield US$1.12 or C$1.15. Remember, this is not the value of the real estate but the ecosystem services produced by that real estate. To put this in context, between 1990 and 2010 about 3% of the world's forests were cut down. Three percent may not sound like a lot, but this is equivalent to 135 million square kilometers. Following deGroot and colleagues' estimates, the value of lost ecosystem services would be between 15.2 and 31.8 trillion int$/year. Using a different resource, fish, mollusk and other aquatic food production increased from 90 tonnes in 2001 to 148 tonnes in 2010; most of this increase has been accomplished through aquaculture, and harvest of wild fish has remained relatively stable.[50] However, more than half of the world's fin fish fisheries are at their maximum sustainable limit and 28% are overexploited, depleted, or recovering from over fishing. Already 40% of the planet's land is in agriculture, and much of what is left simply is not arable. As the human population continues to grow, the pressure on all natural resources will increase. Conservation of natural resources such as wildlife and forests is already facing stiff competition with the desire for these resources. In an attempt to reach a common playing ground in this competition, conservationists try to develop an understanding of real values of the services provided by intact natural resources. With this understanding, agencies can confront those who wish to develop in ecologically sensitive areas with cost/benefit arguments based on valid information.[51,52] Of course there are aesthetic values which defy a dollar figure, but at least an economic price tag is something that even developers can understand.

Establishing values for ecosystem services can also assist agencies and non-government organizations in their efforts to restore or replace conservation lands and resources. The US federal government, for example, has a powerful tool called *Natural Resource Damage Assessment* (NRDA) to determine the value of resource loss due to accidents and corporate abuse. Depending on the area affected, teams from the National Oceanic and Atmospheric Administration (oceans), US Fish and Wildlife Service (wildlife and fisheries) or the Environmental Protection Agency (other damage caused by contaminants or to wetlands) assess the dollar value of loss and try to determine the responsible party who they can sue to recoup those losses. Environment Canada and the Canadian Assessment Agency do the same for that nation. For example, the Deepwater Horizons oil spill that occurred in 2010 resulted in the largest NRDA investigation up to that time; final settlement will take years,[53] and yield over a billion dollars in fines. In this case the FWS evaluated the onshore loss of nesting sites, habitat, and wildlife, while the NOAA monitored and evaluated the effects of the spill on marine life, water quality, sediment loading and other factors tied to the sea.

Another use of determining ecosystem values is in efforts to restore or create new habitat. The Canadian Wildlife Service, Environment Canada, US Forest Service, Fish and Wildlife Service, Bureau of Land Management, and Natural Resources Conservation Service supervise programs to assist states, provinces and landowners to increase conservation efforts on their properties. States and provinces also have support and penalty programs to improve or establish wetlands and erosion control, and to enhance wildlife habitat on private lands. These conservation measures seek to embellish ecosystem services. Programs can provide incentives or penalties. As one example, the US Department of Agriculture has a set of programs that provide incentives to landowners to implement erosion control, develop forests or install wetlands on their properties; they also have programs that penalize landowners for plowing up prairies or draining wetlands and swamps (see Chapter 7). A natural social aspect of these laws is that landowners are much more supportive of incentives and are more likely to follow through in conservation measures than they are with penalty programs.[54]

Another social economic aspect is that agricultural producers are in business to make money. Thus, there is a dynamic between the value of the commodities (corn, soy beans) and the willingness to accept price supports to keep farmland out of production. Conservation in this context needs to be at least partially based on economics, in what is called *market-based incentives*.[55] If the value of corn, for example, is low, producers are more apt to enroll their land into a conservation practice. As commodity prices increase, there is a tipping point when producers will receive more from growing corn, even on marginal land, than not. At this point, enrollment of acreage in conservation programs can decline and existing land may be taken out of conservation. When the value of corn jumped around 2012, the number of acres enrolled in conservation dropped in the United States by about 8% compared to when corn prices were lower.[56] As an added complexity, the chief cause for the high price of corn was its use in the production of ethanol as a way to reduce petroleum consumption. There are many more examples of the interrelationship between conservation of natural resources, social elements and economics that go into sustainability of ecosystem services, but the bottom line is that conservation agencies and NGOs are becoming increasingly aware of the social and economic drivers in their efforts to enhance conservation.

Ecosystem Resilience and Resistance

Two elements associated with ecosystem services are *resilience* and *resistance*. Actually, the terms are characteristics of the ecosystem and can be applied without referring to human needs, but their most common usage is in regard to ecosystem values. Ecological resilience is the amount of perturbation or disturbance an ecosystem can absorb without transitioning to an alternate state or condition.[57,58] Resistance is associated with resilience, but is the ability of an ecosystem to withstand stress without changing. Both contribute to the stability of an ecosystem. By analogy, the factors can be likened to balls. An old tennis ball is pliable and can easily be pushed in, but afterwards it quickly returns to its original shape. This ball would have high resilience but low resistance. A hard rubber ball demonstrates both resilience and resistance in that it takes quite a bit of force to make an indentation but it quickly returns to its original form. Both the rubber ball and the tennis ball are stable. In contrast, a hollow ball made of aluminum foil would have very little resistance or resiliency; it is easy to mold it into various shapes, and when done it stays in its altered form. The aluminum ball cannot be considered stable in this analogy. In a similar way, grassland ecosystems tend to have greater resiliency but less resistance than forests because if a stressor such as fire comes through the grassland will burn quickly but will rebound, often within the same growing period, into a another grassland. A forest will require a hotter, more prolonged fire to burn completely, but when it does it may be decades before it returns to a similar seral stage.

Ecosystem services are connected to resilience and resistance when humans wish to maintain or alter the services provided by a patch of land. Since productive ecosystems are constantly undergoing perturbations, their resistance and resiliency will determine how much active management must be done to maintain or improve the services they provide. However, strong resistance or high resiliency can also work against conservation efforts. For example, we might consider it desirable to convert deserts to agricultural lands, but such conversion would require extensive water, management, and cost to overcome the natural resistance.

Species diversity can be very important to stability, and, while numerous studies have found positive correlations between general species diversity and stability,[59–61] some forms of diversity are better than others. Zell and Hubbart[58] explained that functional and response diversity are key in determining resilience. *Functional diversity* occurs when multiple species are capable of providing the same ecological services. For example, songbirds consume vast amounts of insects, many of which are deleterious to economically useful plants. If there are multiple species of insect-eating songbirds in a forest, the function of depredating insects can still occur even if one species of bird is absent. *Response diversity* occurs when species are able to adapt quickly to perturbations in the ecosystems. Birds with a broad diet range, for example, may be able to tolerate the absence of an insect species by increasing their feeding rates on other insect species.

Biggs and colleagues[62] listed several principles that agencies can use to facilitate enhancing and maintaining resiliency or resistance in an ecosystem. Their first principle is to maintain diversity and redundancy, for the reasons cited above. The second is to manage connectivity, which is the way and degree to which resources, species or social factors disperse, migrate or interact across ecological and social landscapes that include groups, habitats or patches. Connectivity promotes the exchange of information required for the functioning of ecosystems. Strong connectivity spreads out and dilutes disturbances over a wider range of the ecosystem and decreases the risk that one component will be overly taxed. As an example, the authors cited fragmentation of coral reefs. If there are linkages among the remaining stands of coral, recolonization will be facilitated. Contrarily, if the remaining coral outcroppings are scattered or in a position such that currents do not facilitate movement of the mobile larvae, recolonization could be hampered. It is well known that forest fragmentation has been very detrimental to songbirds, in large part due to the loss of connectivity.[63,64] Third, agencies should manage slow variables and feedbacks. Slow is good for stability. Soil composition and phosphorus concentrations in sediment are examples of factors that move slowly in the environment. Negative (retarding) or positive (accelerating) feedback mechanisms can occur when a change in a process occurs. To maintain ecosystem services, these feedback mechanisms may have to be altered to their original condition; conversely, feedback mechanisms can be managed to encourage change.

EVALUATING THE SUCCESS OF CONSERVATION EFFORTS

Given that conservation efforts can be costly, it makes sense that agencies would want to assure that their efforts were effective and efficient. This involves monitoring the effects produced by the project. Monitoring efforts can also increase knowledge about the effectiveness of certain activities, and increase the efficiency of similar projects in the future. Unfortunately, the type of post-project monitoring that is required to make these assessments is seldom carried out. Administrators at both the state (provincial) and federal levels too often provide funds for implementation of a project, but not for post-implementation evaluation. Rather, they are eager to get on to the next project and reluctant to spend money or personnel on evaluation. Evaluations can be costly and difficult, they may generate unwanted input, and their recommendations may be difficult to implement. Nevertheless, evaluation is critical to learning, to improving the effectiveness and efficiency of programs, and to directing change.[65]

Several methods of monitoring and evaluation have been developed, each with its own advantages and disadvantages.[66] Stem et al.[35] distinguished methods for status assessment from those that measured effectiveness of a conservation project. Status assessment is an older but still effective approach towards conservation evaluation. A common method of status assessment is estimating the abundance of individuals in a population through time. The question here is generally expressed as: Did conservation practice Y have an effect on population X and, if it did, it what direction did it affect X? The question of "how much" it affected population X (that is, actual changes in numbers) is a more costly endeavor and may not be necessary.

About 30 years ago a quicker method, rapid bioassessment, was developed. The general concept has been extensively used and formalized in stream and wetland assessments,[67,68] but it is also useful in terrestrial landscapes.[69] In this process, a series of well-conceived questions about an area is developed to obtain considerable information in a short period of time — perhaps hours to a day. The method is relatively inexpensive compared to detailed analyses and, if conducted with consistency, can provide insight to changes within a conservation area or population through time. This approach has been enhanced and widely used by NGOs such as Conservation International and the World Wildlife Fund.[35] Rapid assessments, by their very nature and lack of rigor, are open to subjectivity and reduced reliability compared to some other methods of evaluation.

Global or regional assessments use macro-indicators (estimates of species diversity, climate variables, major social activities, etc.) to create wide-scale interpretations of an area on an infrequent (5- or 10-year) basis. One of the more ambitious of these assessment programs was the Millennium Assessment, in 2001, which involved over 1000 scientists from around the world and provided a valuable benchmark for future assessments.[70] *National Report Cards*, published annually by several NGOs, provide a serial documentation of the state of the region, species, or world, but on a limited scale.

A more accurate form of evaluation, described by Stem *et al.*,[66] quantifies effectiveness of specific conservation programs. These may include one-time assessments, called *impact assessments* or statements, or a continual evaluation approach, referred to as *adaptive management* (AM). Adaptive management has been around since the 1960s and was formalized in the 1970s,[71] well before Soulé's seminal paper on the meaning of conservation.[11] It is an iterative process that allows biologists to make repeated assessments and refinements in their plans as the implementation process is occurring.[72] I like the definition for adaptive management provided by Allen *et al.*[38] (p. 1339), because it emphasizes that managers do not always know what they are doing, despite what they might like to think: "Adaptive management is an approach to natural resource management based on the philosophy that knowledge is incomplete and much of what we think we know is actually wrong, but despite uncertainty managers and policy makers must act." Although it can serve as a very effective means to reduce the uncertainty and inefficiency associated with trial-and-error forms of management, AM is not used as commonly as it should be, just like the other methods of evaluation. The process has some similarities to the scientific method (Figure 1.9), and has sometimes been called managerial research. Starting at the top of the diagram, a manager defines the problem, which is comparable to stating a hypothesis. Unlike the scientific method, however, adaptive management has desired goals or outcomes which are clearly defined in the next few steps. These goals can be likened to "alternative hypotheses" in experiments. Implementation of the program is analogous to conducting the experiment, and the next few steps of monitoring, evaluating, adjusting and identifying the next step are all akin to research studies.

Adaptive management is appropriate for many natural resource issues. It has been used in cases involving species and habitat declines, prescribed burning and wildlife, water basin management, and forest management. Perhaps the most common application of AM, however, is in waterfowl harvest management. The FWS has used AM or, as it is sometimes called, *adaptive harvest management* on mallards (*Anas platyrhynchos*) and other waterfowl since 1995.[73] The emphasis on using AM for waterfowl management has been become an important area of cooperation between the FWS and USGS in the Department of the Interior,[74] where it has led to the publication of a national protocol of adaptive management.[75] Adaptive management has been incorporated into several Canadian agencies, and is an important part of the development of Environmental Assessments by the Canadian Environmental Assessment Agency.[76]

FIGURE 1.9   A diagram illustrating the steps in adaptive management.

Study Questions

1.1. What is the difference between conservation and preservation? Under what general conditions might either be the appropriate course of action?

1.2. Why is conservation increasingly being considered in terms of the relationship between human needs and the environment?

1.3. What is the difference between conservation biology and conservation science?

1.4. Describe some of the ecosystem services provided by a particular type of ecosystem, such as a wetland, lake, prairie or forest.

1.5. Discuss the differences between the concepts of sustained use and sustainability.

1.6. Discuss the steps of adaptive management. How are these steps similar to conducting an experiment?

REFERENCES

1. Père David's Deer (*Elaphurus davidianus*). <http://www.arkive.org/pere-davids-deer/ela-phurus-davidianus/>.
2. Père David's Deer. <http://en.wikipedia.org/wiki/P%C3%A8re_David%27s_deer>.
3. US Fish and Wildlife Service. Biology of the California Condor. <http://www.fws.gov/hop-permountain/cacorecoveryprogram/CACO%20Biology.html>.
4. Collins PW, Snyder NFR, Emslie SD. Faunal remains in California Condor nest caves. *Condor*. 2000;102:222–227.
5. Church ME, Gwiazda R, Risebrough RW, et al. Ammunition is the principal source of lead accumulated by California Condors re-introduced to the wild. *Environ Sci Technol*. 2006;40:6143–6150.

6. California Condor Recovery Program. <http://www.fws.gov/hoppermountain/cacorecoveryprogram/CACO%20Biology.html>.
7. Conservation efforts and current status of the endangered California Condor. Defenders of Wildlife – California Program. <https://www.defenders.org/sites/default/files/publications/california_condor_presentation.pdf>.
8. Green Earth Inc. <http://www.greenearthinc.org/>.
9. The Nature Conservancy. <http://www.nature.org>.
10. Shultis J, Moore T. American and Canadian national park agency responses to declining visitations. *J Leisure Res*. 2011;43:110−132.
11. Monz CA, Cole DN, Leung YF, Marion JL. Sustaining visitor use in protected areas: future opportunities in recreation ecology research based on the USA experience. *Environ Manage*. 2010;45:551−562.
12. Minteer BA, Corley EA. Conservation or preservation? A qualitative study of the conceptual foundations of natural resource management. *J Ag Environ Ethics*. 2007;20:307−333.
13. USDA Agricultural Research Center. <http://www.ars.usda.gov/main/site_main.htm?modecode=54-02-05-00>.
14. McCleery DW. *American Forests: A History of Resiliency and Recovery*. Collingdale, PA: Diane Publishing Co; 1993.
15. Norton BG. Conservation and preservation: a conceptual rehabilitation. Environ Ethics. 8:195−220.
16. Soulé M. What is conservation biology? *BioScience*. 1985;35:727−734.
17. Soulé M. Land-use planning and wildlife maintenance – guidelines for conserving wildlife in an urban landscape. *J Am Plan Assoc*. 1991;57:313−323.
18. Soulé M, Alberts AC, Bolger DT. The effects of habitat fragmentation on chaparral plants and vertebrates. *Oikos*. 1992;63:39−47.
19. Mills LS, Soulé ME, Doak DF. The keystone-species concept in ecology and conservation. *Bioscience*. 1993;43:219−224.
20. Sanderson EW, Redford KH, Weber B, et al. The ecological future of the North American Bison: conceiving long-term, large-scale conservation of wildlife. *Conserv Biol*. 2008;22:252−266.
21. Pearce JL, Kirk DA, Lane CP, et al. Prioritizing avian conservation areas for the Yellowstone to Yukon region of North America. *Biol Conserv*. 2008;141:908−924.
22. Collard RC. Cougar−human entanglements and the biopolitical un/making of safe space. *Environ Planning D-Soc Space*. 2012;30:23−42.
23. Clevenger AP. Conservation value of wildlife crossings: measures of performance and research directions. *Gaia-Ecol Perspect Sci and Soc*. 2005;14:124−129.
24. Kroger R, Dunne EJ, Novak J, et al. Downstream approaches to phosphorus management in agricultural landscapes: regional applicability and use. *Sci Total Environ*. 2013;442:263−274.
25. Keough HL, Blahna DJ. Achieving integrative, collaborative ecosystem management. *Conserv Biol*. 2006;20:1373−1382.
26. Garriga N, Santos X, Montroi A, et al. Are protected areas truly protected? the impact of road traffic on vertebrate fauna. *Biodiv Conserv*. 2012;21:2761−2774.
27. Kim MK, Daigle JJ. Monitoring of vegetation impact due to trampling on Cadillac Mountain Summit using high spatial resolution remote sensing data sets. *Environ Manage*. 2012;50:956−968.
28. Meunier G, Lavoie C. Roads as corridors for invasive plant species: new evidence form smooth bedstraw (*Galium mollugo*). *Invasive Plant Sci Manage*. 2012;5:92−100.

29. Van Marwijk RBM, Elands BHM, Kampen JK, et al. Public perceptions of the attractiveness of restored nature. *Restor Ecol.* 2012;20:773−780.

30. Chavez DJ. Natural areas and urban populations: communication and environmental education challenges and actions in outdoor recreation. *J Forest.* 2005;103:407−410.

31. Kareiva P, Marvier M. What is conservation science? *BioScience.* 2012;62:962−969.

32. US National Park Service Yellowstone National Park. <http://yellowstone.net/history/timeline/the-pre-park-years-1795-1871>.

33. Springer J, Campese J, Painter M. *Conservation and Human Rights: Key issues and Contexts. Scoping Paper for the Conservation Initiative for Human Rights.* Washington DC: World Wildlife Fund; 2011:1−45. <http://assets.worldwildlife.org/publications/364/files/original/16_Conservation_and_Human_Rights_Key_Issues_and_Contexts_.pdf?1345736627>.

34. Millennium Declaration. <http://www.un.org/millennium/declaration/ares552e.htm>.

35. World Summit for Social Development. <http://www.un.org/esa/socdev/wssd/textversion/>.

36. Cato MS. *Green Economics: An Introduction to Theory, Policy and Practice.* London, UK: Earthscan; 2009.

37. Adams, WM. The Future of Sustainability: Re-thinking Environment and Development in the Twenty-first Century. Report of the IUCN Renowned Thinkers Meeting, 29−31 January 2006; 2006.

38. Becker CU. *Sustainability Ethics and Sustainability Research.* New York, NY: Springer; 2012.

39. Web of Science. <http://apps.webofknowledge.com>.

40. Sharifi A, Murayama A. A critical review of seven selected neighborhood sustainability assessment tools. *Environ Impact Assess Rev.* 2013;38:73−87.

41. Sullivan E, Ward PM. Sustainable housing applications and policies for low-income self-build and housing rehab. *Habitat Intern.* 2012;36:312−323.

42. Iacovidou E, Ohandja D-G, Gronow J, Voulvoulis N. The household use of food waste disposal units as a waste management option: a review. *Crit Rev Environ Sci Technol.* 2012;42:1485−1508.

43. Farmer MC, Wallace MC, Shiroya M. Bird diversity indicates ecological value in urban home prices. *Urban Ecosyst.* 2013;16:131−144.

44. US Fish and Wildlife Service 2012. *National Survey of Fishing, Hunting and Wildlife-Associated Recreation: National Overview.* Washington, DC: US FWS; 2011.

45. Betts AK, Ball JH. Albedo over the boreal forest. *J Geophys Res Atmos.* 1997;102:28901−28909.

46. Hackett KJ. Bee benefits to agriculture. *Ag Res.* 2004;:2.

47. United Nations Department of Economic and Social Affairs.World Population Prospects, the 2010 Revision. New York, NY (available at <http://esa.un.org/wpp/Excel-Data/population.htm)>; 2010.

48. Costanza d'Arge R, deGroot R, Farber R, et al. The value of the world's ecosystem services and natural capital. *Nature.* 1997;387:253−260.

49. deGroot R, Brander L, van der Ploeg L, et al. Global estimates of the value of ecosystems and their services in monetary units. *Ecosys Serv.* 2012;1:50−61.

50. FAO (Food and Agriculture Organization of the United Nations). Fisheries and Aquaculture Department Statistics (available at <http://www.fao.org/fishery/statistics/en)>; 2010

51. Wainger L, Mazzotta M. Realizing the potential ecosystem services: a framework for relating ecological changes to economic benefits. *Environ Manage.* 2011;48:710−733.

52. Ernston H. The social production of ecosystem services: a framework for studying environmental justice and ecological complexity in urbanized landscapes. *Landscape Urban Plan.* 2013;109:7–17.

53. National Oceanic and Atmospheric Administration. <http://www.gulfspillrestoration.noaa.gov/2012/04/status-update-on-nrda/>.

54. Lant CL, Kraft SD, Gillman KR. The 1990 farm bill and water quality in Corn Belt watersheds: conserving remaining wetlands and restoring farmed wetlands. *J Soil Water Conserv.* 1995;50:201–205.

55. Lockie S. Market instruments, ecosystem services and property rights: assumptions and conditions for sustained social and ecological benefits. *Land Use Policy.* 2013;31:90–98.

56. US Department of Agriculture. <http://www.fsa.usda.gov/FSA/webapp?area=home&subject=copr&topic=crp-st>.

57. Holling C. Resilience and stability of ecological systems. *Ann Rev Ecol System.* 1973;4:1–23.

58. Zell C, Hubbart JA. Interdisciplinary linkages of biophysical processes and resilience theory: pursuing predictability. *Ecol Model.* 2013;248:1–10.

59. MacArthur R. Fluctuations of animal populations and a measure of community stability. *Ecology.* 1955;36:533–536.

60. Ives A, Carpenter S. Stability and diversity of ecosystems. *Science.* 2007;317:58–62.

61. May R. Will a large complex system be stable? *Nature.* 1972;238:413–414.

62. Biggs R, Schlüter M, Biggs D, et al. Toward principles for enhancing the resilience of ecosystem services. *Ann Rev Environ Resour.* 2012;37:421–448.

63. Melles S, Fortin MJ, Badzinski D, Lindsay K. Relative importance of nesting habitat and measures of connectivity in predicting the occurrence of forest songbird in fragmented landscapes. *Avian Conserv Ecol.* 2012;7:3.

64. Martensen AC, Riberio MC, Banks-Leite C, et al. Associations of forest cover, fragment area, and connectivity with neotropical understory bird species richness and abundance. *Conserv Biol.* 2012;26:1100–1111.

65. Kleiman DG, Reading RP, Miller BJ, et al. Improving the evaluation of conservation programs. *Conserv Biol.* 2000;14:356–365.

66. Stem C, Morgouluis R, Slafsky N, Brown M. Monitoring and evaluation in conservation: a review of trends and approaches. *Conserv Biol.* 2005;19:295–309.

67. Barbour MT, Plafkin JL, Bradley BP, et al. Evaluation of EPA's rapid bioassessment benthic metrics – metric redundancy and variability among reference stream sites. *Environ Toxicol Chem.* 1992;11:437–449.

68. Ainslie WB. Rapid wetland functional assessment – its role and utility in the regulatory arena. *Water Air Soil Poll.* 1994;77:433–444.

69. MacLeod R, Herzog SK, Maccormick A, *et al.* Rapid monitoring of species abundance for biodiversity conservation: consistency and reliability of the MacKinnon lists technique. Biol Conserv 144:1374–1381.

70. Powledge F. The millennium assessment. *BioScience.* 2006;56:880–886.

71. Holling CS. *Adaptive Environmental Assessment and Management.* London, UK: John Wiley & Sons; 1978.

72. Allen CR, Fontaine JJ, Pope KL, Garmestani A. Adaptive management for a turbulent future; 2011.

73. Nichols JD, Runge MC, Johnson FA, Williams BK. Adaptive harvest management of North American waterfowl populations: a brief history and future prospects. *J Ornithol.* 2007;148(suppl 2):S343–S349.

74. Moore CT, Lonsdorf EV, Knutson MG, et al. Adaptive management in the US National Wildlife Refuge system: science–management partnerships for conservation delivery. *J Environ Manage*. 2011;92:1395–1402.

75. Williams BK, Szaro RC, Shapiro CD. *Adaptive management: The US Department of the Interior Technical Guide*. Washington, DC: Adaptive Management Working Group, US Department of the Interior; 2009.

76. Operational Policy Statement, Canadian Environmental Assessment Agency. <http://www.ceaa-acee.gc.ca/default.asp?lang=En&n=50139251-1>.

Chapter 2

History of Wildlife and Natural Resource Conservation

Terms to Know

- Smallpox
- Myth of Superabundance
- Protectionism
- Manifest Destiny
- Homestead Acts
- Voyageurs
- Kaibab Plateau
- *Game Management* (1933)
- Environmentalism
- *Silent Spring* (1962)
- *The Population Bomb* (1968)
- *A Sand County Almanac* (1940)

INTRODUCTION

The philosopher George Santayana (1905)[1] wrote: "Those who cannot remember the past are condemned to repeat it." The history of use and abuse of natural resources in North America can be characterized by a series of mistakes, poor judgment and greed mixed with a gradual learning process on the meaning and application of conservation that continues through today. This history starts with pre-settlement by Europeans, and extends through an early period of colonization to a rather rapid movement into all parts of a developing continent. It has evolved from a fear of the wilderness, to attempts to subjugate nature to human greed, to an eventual understanding of the value of maintaining our natural resources for current and future generations. By no means have we arrived at perfection in our management efforts, and there are many, many competing interests all with their own perspectives on conservation, but we have made substantial progress over the past centuries. To a large extent, the events or periods of conservation that occurred in the United States

D.W. Sparling: Natural Resources Administration. DOI: http://dx.doi.org/10.1016/B978-0-12-404647-4.00002-7

27

also occurred in Canada but about 20 to 50 years later. Let us recall this history so as to better understand where we are now in terms of natural resource conservation.

THE PRE-EUROPEAN ERA (PRIOR TO THE 1500s)

The origin of the first humans in North America is hotly debated. The traditional explanation is that they probably crossed a land bridge across the Bering Sea some 12,000 years ago. An alternative explanation is that a group of travelers from Australia crossed the Pacific Ocean about 13,500 years ago and initially settled in Baja, California.[2] Regardless of the origin of the first settlers, both groups gradually moved out into the rest of North and South America.

The influence of northern hemisphere Native Americans or Amerindians on the landscape has also been contested, but most likely it varied considerably depending on location. It is unlikely that the romantic notion of the "noble savage", as described by Rousseau and others, was very common, if indeed it existed at all. Pre-Columbian Amerindians lived in a variety of conditions ranging from nomadic hunters and gatherers, to small villages, to a few great civilizations consisting of thousands of people living in what can only be called cities. Clearly, the need for resources — timber for wood; plants and animals for fiber, food and labor; and minerals and metals for tools and jewelry — varied with human population density and lifestyle. Prior to Columbus's fateful landing at San Salvador, the indigenous population in the Americas was probably around 40−80 million people although only 3.8 million lived in what would become the United States and Canada.[3] Thus, while sections of Mexico, Central and South America sustained sizable populations, most of the area north of the Rio Grande was very sparsely populated. The only known northern "city" in Pre-Colombian times was Cahokia, in what is currently Illinois (Figure 2.1). Over the past few decades archeologists have discovered that this city was home for as many as 60,000 people during its peak in the 1200s. However, it remained the largest city north of the Rio Grande until 1775, when it was surpassed by Philadelphia and New York City.[4] Farming apparently was the primary means of acquiring food.

Of the other native Americans north of the Rio Grande, most lived in villages or hamlets along the East Coast. Other communities sprang up along major rivers, and there were populations in California and the Pacific Northwest. Human populations were very sparse between the Mississippi and the Rocky Mountains and, with the exception of the Hopis and Pueblos, most of these were nomadic tribes that sustained themselves through hunting and gathering. Where population density was higher, there is little doubt that Amerindians impacted the local environment. In New England and the Midwest, substantial timber cutting had occurred well before European colonization. Rather than dense climax forests, much of the woods had been

FIGURE 2.1   Artist's rendition of Cahokia, a pre-Columbian city of 40,000 inhabitants along the Mississippi River in Illinois. *Reproduced with permission of Cahokia Mounds Historic Site.*

cut to make way for agriculture. Forests that were not totally clear cut had open canopies and sparse understories.[5] The openings that were produced were actually beneficial for some species of wildlife. Outside of wooded areas, the predominant habitat was semi-permanent grasslands and open fallow ground. Because some Amerindians had developed sustainable agriculture with rotation of crops and fertilization, villages could exist in one location for many years.

One of the major factors shaping the environment at this time was fire, either naturally set or of human origin. Amerindians in the East may have used fire extensively to suppress woody growth around inhabitations.[6] However, Denevan (2003) has disagreed, saying that there is not good evidence for this.[3] The case that the eastern edge of the prairie/forest continuum was greatly affected by fires set by indigenous peoples presents a stronger argument.[7] Fires set by Amerindians in the Midwest and other areas of the continent, coupled with climatic conditions, altered the prairie/forest ecotone by hundreds of miles (Figure 2.2).[8]

The effect of Amerindians on wildlife populations is controversial. Some researchers suggest that early indigenous peoples (i.e., more than 10,000 years ago) decimated local populations of mammoths, camels, elk, deer, and other species (see, for example, Martin 1984,[9] Grayson and Meltzer 2003,[10] Forkey 2012,[11] Crosby 1976[12]). Others suggest that the extinctions occurred too rapidly to be due to the relatively low human populations at the time, and that other factors such as climate change may be more important in these extinctions and population declines (e.g., Grayson and Meltzer 2003[10]).

In summary, before the arrival of Europeans, human populations were sparse, and concentrated in a few areas where they may have had some effect on regional resources. However, indigenous peoples may have influenced the range of prairies and forests in the Midwest, the Pacific

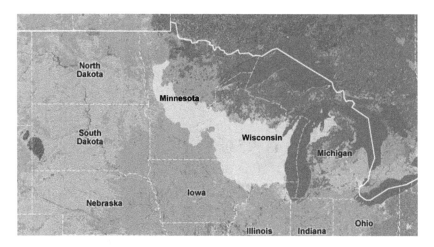

FIGURE 2.2   Midwest savanna ecotone – one of the several regions in the Midwest affected by fire and climate. *Credit: World Wildlife Fund, Ecological Encyclopedia. This figure is reproduced in color in the color plate section.*

Northwest, and elsewhere and may have exerted pressure on large herbivores, at least locally. Once Europeans arrived, however, the influence of Amerindians declined as their populations were severely decimated.

Along with other things, Europeans brought with them numerous diseases to which the natives had no immunity. The first diseases came with Columbus (Figure 2.3), but their dispersion increased as more Europeans came to the New World. Smallpox, cholera, influenza, pneumonia, plague, and whooping cough wiped out entire villages and decimated regional populations.[11] While the deaths began with Columbus' expeditions, they continued for centuries afterwards. For example, a plague, possibly bubonic, swept coastal New England from Cape Cod to Southern Maine from 1616 to 1619, killing up to 90% of those infected.[12] While Europeans had some resistance to these diseases, American Indians did not, and most epidemics were far more severe for them. During the 1630s and 1640s, for example, smallpox ran rampant through the St Lawrence River, Great Lakes region, killing more than half of the Huron and Iroquois nations. Smallpox continued to wipe out Amerindians for over 100 years, with major epidemics in 1738 in the Cherokees, 1759 in the Catawbas, and a few years later in the Piegan tribe. Later, smallpox decimated the plains tribes and nearly half of the natives in the Great Plains were killed by epidemics.[11]

These epidemics profoundly affected the landscape of North America prior to intense European colonization. A great many of the villages and tribes in eastern North America were wiped out in the 1500s and early 1600s, so that by the time Puritans and Pilgrims settled, the open canopy and park-like environment of New England had been replaced by dense overgrowth and forests.

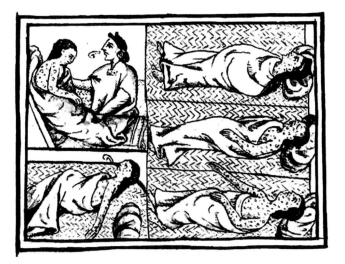

FIGURE 2.3   An Aztec drawing of Amerindians infected with smallpox, and dying.

EUROPEAN INCURSIONS AND EARLY SETTLEMENT (LATE 1500s TO 1700)

Explorers and colonists from several western European countries came to the New World for a variety of reasons. However, three nations, Spain, France and England, predominated. The explorations were funded by the nobility – most often the kings and queens of these countries – and their objectives differed decidedly according to national interests. The primary goal of Spanish explorers was to find riches such as gold and precious gems. Of secondary interest, missionaries were sent with the expeditions to Christianize the pagans. The Spanish were the great explorers who journeyed into South, Central, and southwestern North America, and Florida (Figure 2.4). Although the city of St Augustine in Florida was founded 80 years before the Pilgrims arrived, the Spanish were not major colonists in North America. The French were primarily interested in furs. They explored much of the Mississippi River drainage and were very active in Canada along the St Lawrence River and the Pacific Northwest, where the pelts were of high quality. Along with the fur trade, the French established trading posts that eventually became great cities such as Montreal, St Louis, and New Orleans. They constructed settlements on Caribbean Islands like St Martinique, St Croix and St Lucia. The French also brought Jesuit missionaries with them to convert natives. England was most successful in establishing colonies in what would become North America, and contributed substantially to the settlement of the Mid-Atlantic and New England regions.

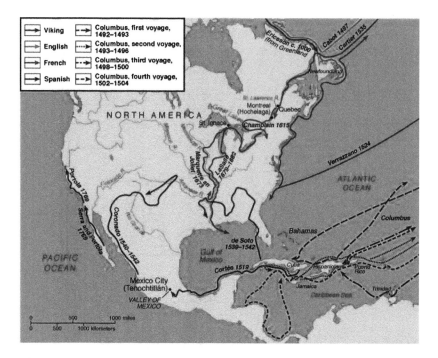

FIGURE 2.4   Early exploration routes of Columbus, the Spanish and the French. *This figure is reproduced in color in the color plate section.*

People came to America for a variety of reasons. The Pilgrims, Puritans and Catholics came seeking religious freedom. The concept of primogenitor still prevailed in much of Europe, meaning that the firstborn son inherited most or all of the family estate. Therefore, many of the early colonists were second, third or fourth sons of aristocracy who came to establish a new life. Unpaid debt and criminal actions were subject to long prison terms, so many others came to North America to escape the gaols and dungeons of the time. Several of the English settlements were started by investment groups funding excursions, such as the Company of Merchant Investors for the Pilgrims, or the Massachusetts Bay Company for the Puritans and New England. Another form of settlement was through charters given by the kings to loyal subjects. For example, King Charles I of England granted Lord Baltimore a charter for founding Maryland, and Charles II established a charter for the Carolinas to the Lords Proprietors. Agents needed to recruit people who were willing to travel thousands of miles to a strange and foreign land for the rest of their lives, and they often could not be too finicky about whom they selected; it was seldom the elite or the highly skilled who wanted to make the trip. Within 150 years of the arrival of Columbus, most of the northern Atlantic seaboard had been settled and much of the West had been explored (Figure 2.5).

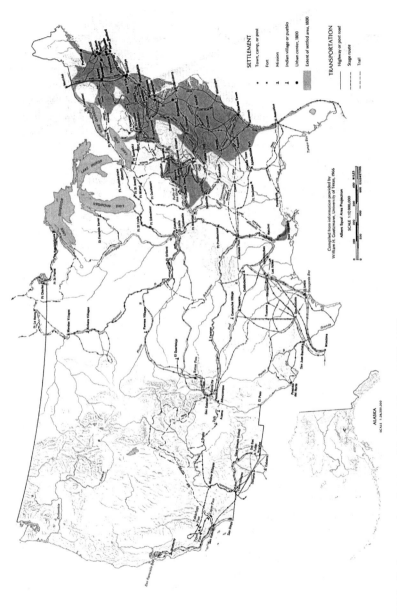

FIGURE 2.5 Settlement and exploration of the United States, 1685–1800.

What was the attitude of the first colonists to the New World? As the saying goes, "try to place yourself in their moccasins". You had probably grown up in a city such as London or, if you were lucky, in very well-established countryside. The King or his designated lords held title to all real estate back in England, and to all the wildlife that lived there. Most likely you were of common stock, and if you were skilled in any crafts, you were among the minority. So, in order to survive, you had to carve out a humble farmstead in the vast wilderness around you. Clearing the land, building a house, plowing fields, planting and hoeing were hard work. As you labored you had to be constantly on the watch for hostile natives or wild predators such as wolves, bears and panthers. If they did not get you, there were also the deer, birds and other herbivores who found your garden patch delectable. You did not dare venture far from town for fear that the wilderness would swallow you up. Those not from a farming background were not experienced in techniques such as fertilization, and after a few years the soil could be depleted of nutrients. Because the crop seeds you brought with you often did not do as well as they did back in England, you had the added concern of wondering if your meager crops would carry you through the winter. The first years of being a colonist were not easy, to say the least. As a result, the earliest colonists did not venture far beyond their little piece of civilization and had a relatively minor impact on the natural resources of their regions.

After about a generation, colonists began to feel more comfortable in their surroundings. In some areas, the Amerindians had either proven friendly or were chased out by musket and militia. People began to explore further and further away from their villages, and to realize just how rich the natural resources were. By 1700 the East Coast of North America had been sectioned into specific colonies, most of which were under British control. As the New World became more hospitable it attracted a greater variety of colonists, who built up significantly the major cities of Boston, Philadelphia, New York and Baltimore.

DAWNING OF THE MYTH OF SUPERABUNDANCE (1700 TO THE 1850s)

As the European-based population in North America grew, more and more people moved out into the frontier. The Appalachian Mountains and international treaties (France claimed much of the land west of the Appalachians, and Spain claimed a good section of the Southeast) provided an obstacle that stymied too rapid a western expansion, so human movement also extended south into Georgia and northern Florida, although the Spanish blocked progress in that direction. As awareness of the incredible wealth of natural resources grew, a general *Myth of Superabundance* (the term coined by Stewart Udall, Secretary of the Interior, in 1963[13]) became rampant throughout the nations. "The natural resources of this country are so huge and so widespread that they never can be exhausted", people argued. As a result, there was great wastage of wildlife, forests, lands and waters. The first animals to go were the large predators. These became man's enemies, for they threatened both human life and economic loss through depredation on livestock. The first bounties on wolves were set by Connecticut in 1647. By the time the Revolutionary War started, almost all large predators had been eliminated from the settled areas of the East.[14] However, predators were not the only group to be decimated during this period.

Passenger pigeons (*Ectopistes migratorius*), along with many other game animals, were caught or killed *en masse*, packed into barrels and shipped to the major markets in Boston, New York and Philadelphia. Taber and Payne (2003)[15] described a typical large city meat market as displaying 2 species of swans, 4 species of geese, 26 species of ducks, 6 of grouse; 43 of marsh birds (including gulls, horned grebe and coots), turkeys, quail, woodcock, passenger pigeons, songbirds, bison, deer, elk, moose, caribou, antelope, bighorn sheep, porcupine, bear, bobcat, raccoon, possum, skunk, muskrat and beaver. Many of the animals were sold for less than a dollar per dozen or a few cents per pound. An inventory in Chicago during 1873 included 300,000 quail at $1.25/dozen, 600,000 prairie chickens at $3.25/dozen, 450,000 pounds of venison at $0.08/lb, 400,000 pounds of bison at $0.07/lb, and so on.

In 1823 James Fennimore Cooper reported on a massive slaughter of passenger pigeons, perhaps the most numerous species ever to live on Earth. In his novel *The Pioneers*, Cooper described a hunt for passenger pigeons: "None pretended to collect the game, which lay scattered over the fields in such profusion as to cover the very ground with the fluttering victims". Between 1908 and 1813, the ornithologist Alexander Wilson described a personal experience of a passenger pigeon migration while he rode between Shelbyville and Frankfort, Kentucky.[16] A breeding rookery of these birds was reported to have a width of several miles and a length of 40 miles. During one migration, Wilson estimated that the entire body of birds numbered 2,230,272,000! Further, their combined food intake was estimated at 17,424,000 bushels of wild grain and seed per day. Less than 100 years later the last passenger pigeon, Martha, died in the Cincinnati Zoo (Figure 2.6). This incredibly abundant bird was the victim of overhunting, deforestation, and ignorance.

Canada was a home for many of the bearers of fur, so ardently sought by European ladies and gentlemen as early as the late 1500s. French fur trappers or *voyageurs* plied Canadian rivers, trapping or trading for beaver (*Castor canadensis*), mink (*Mustela vison*), marten (*Martes Americana*), lynx (*Lynx lynx*), fox, rabbit, snowshoe hare (*Lepus americanus*) and many other species. As the years progressed, England's power in Canada grew and eventually the French influence was largely restricted to Quebec. The English may have gotten a late start in Canadian settlement, but in 1670 the Hudson's

FIGURE 2.6　Martha, the last passenger pigeon, died in Cincinnati in 1932.

Bay Company was given a charter by King Charles II allowing it exclusive trading in furs in eastern Canada. Some hundred years later the Northwest Company was established to trade in western Canada, giving John Jacob Astor competition until the Northwest Company acquired Astoria. The Northwest Company, in turn, was eventually acquired by the Hudson Bay Company. Beaver was the king of the Canadian fur trade. In the 1700s, beaver pelts were being exported to many parts of Europe and commanding top dollar. Raw skins were exported to England, where they were converted into hats, either as fur or manufactured into beaver wool (Figure 2.7). In 1700, 69,500 beaver hats were exported from England, and almost the same number of felt hats. Sixty years later, slightly over 500,000 beaver hats and 370,000 felt halts were shipped from England. Between 1700 and 1770, more than 21 million beaver and felt hats were exported. England also exported raw pelts to manufacturers in other parts of Europe. During 1760, markets for beaver hats and pelts grew dramatically, such that the value of beaver hat sales to Portugal alone was £89,000 in 1756 to 1760, representing about 300,000 hats, or two-thirds of the entire export trade.[17] It may not be really accurate, but it's interesting to realize that after inflation and conversion of currency, that £89,000 would be worth about $19.3 million today. By the early to mid-1800s, changes in fashion ended the lucrative trade in beavers; however, by then many populations had been extirpated and it would take many years for them to recover.

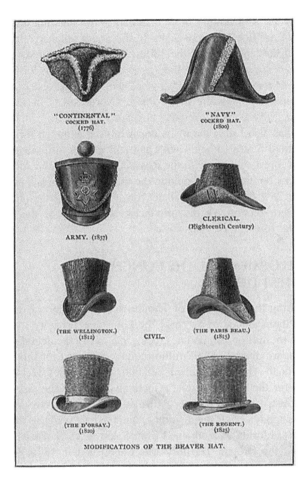

FIGURE 2.7   Different styles of men's beaver hats during the 1700s.

Wanton abuse of natural resources was not limited to wildlife. So many forests were clear-cut that it prompted the English visitor Isaac Weld to write home that Americans apparently hated trees and "cut away all before them without mercy; not one is spared; all share ... in the general havoc" (Taylor 2003,[18] p. 113). By 1800, wood fuel was becoming scarce in large cities. Whether the eastern forests actually represented old growth is arguable, since, as we have seen, Amerindians cut down sizable tracts of forest well before European settlement. However, standing forests contained many huge trees of considerable value. Clear-cutting, in which all marketable trees were removed from an area over a short period of time with no effort to reseed or restore new growth, was the prevailing technique. In the north, clear-cutting started to exceed regeneration of new trees as early as 1700, and by the beginning of the 20th century more than half of the standing timber, 170 million acres (69 million ha) were gone. In the South, intensive logging started later, around the beginning of the 1800s, but over a hundred years southern forests were reduced by 130 million acres (53 million ha), or 37%. Overall, 95% of the forests in the United States have been harvested at least once, with uncut remnants existing in the Pacific Northwest.[19] In addition to timber harvesting, disease and fire have also taken their toll.

Canada has also lost some of its uncut forests, although it had a lot more than the United States to begin with and still retains millions of hectares of unharvested forests. Along with the Russian steppes and the Amazon rainforests, Canada's boreal forest is among the three largest forests in the world. It accounts for 25% of the world's intact forests.[20] Most of the intact forest lies in the northern sections of the provinces; about 40% of the southern old-growth boreal forests have been lost.[21]

THE DEVELOPMENT OF THE CONCEPT OF MANIFEST DESTINY

In 1803, during the presidency of Thomas Jefferson, the new government of the United States agreed to purchase the Louisiana Territory from France for a total of $15 million, or about 3 cents per acre. Incidentally, this would have been approximately $312 million today, corrected for inflation − still a very good deal. The deal gained 828,000 square miles (2,144,000 square kilometers) for the United States, effectively doubling the size of the country. In contrast, Canada became a British possession pretty much with the area we see today. Later acquisitions continued to increase the size of the United States: Florida was obtained from Spain in 1819, Texas was annexed in 1845, and the Oregon Territory was established in 1846; Mexico ceded much of the Southwest in 1848, with the rest coming with the Gadsden Purchase in 1853, the purchase of Alaska in 1867, and the annexation of Hawaii in 1898, which brought the aggregate area of the United States to its present boundaries of 3,718,700 square miles (9,629,575 square kilometers). In comparison, Canada comprises 3,855,000 square miles (9,985,000 square kilometers).

Within 100 years, the United States was faced with some important questions. What should we do with all of this land? What should we do with all of the wealth tied up in natural resources on this land? As the singer Joni Mitchell wrote in *The Big Yellow Taxi* (1970), "you don't know what you've got until it's gone". Blessed by more resources than could be imagined, the waste of natural resources went unabated for many more years. Adding to this situation was a new philosophy termed *Manifest Destiny*. The concept that America was a special nation, particularly blessed by God, was not a new idea − it went back to the framers of the US Constitution − but it took on a new impetus in the mid 1800s. In 1845, newspaper editor John

O'Sullivan proclaimed that America was ordained by God to move into these new lands and subdue them. It was not only our right, it was our duty. This rhetoric spawned fervor to settle the newly acquired territories and exploit the gifts that were put there for us. Canadians joined in on this concept about a decade or two later. Settlers began to move into the territories in droves. The discovery of gold at Sutter's Mill in California in 1848 further spurred movement into the west. Around 50 years later the Klondike or Yukon gold rush occurred, but it did not lead to such a surge in population because the climate kept many people from settling there permanently. It was during this time that many new states were established. In 1862 Congress passed the *Homestead Act*, which gave away 160 acres (65 ha) and eventually 640 acres (260 ha) to anyone establishing a claim, making minimal improvements on territorial lands, and living there for at least five years.

Interest in settlement was slow during the Civil War, but picked up soon after. Many editors picked up Horace Greeley's mantra of "Go West young man", further encouraging people to pack up and seek new lands. The first transcontinental railway was constructed between 1863 and 1869, giving people heightened access to the middle of the nation. The net effect of these combined actions was a rapid expansion of humans into the west. This expansion was not without cost, for the land was already occupied by native Americans; however, since it was believed to be "God's will" that white settlers take over the lands, the prior inhabitants had to go and were eventually sent to various reservations and poor quality lands unwanted by those of European descent.

Movement of settlers into the west increased the slaughter of wildlife and the despoliation of the land. During the building of the transcontinental railways, buffalo hunters killed thousands of bison to feed the workers; after the railroads were completed, the companies encouraged hunters to come and make use of the easy access to game animals, including elk (*Cervus elaphus*), moose (*Alces alces*), bighorn sheep (*Ovis canadensis*), and black tailed and mule deer (*Odocoileus heminous*). In fact, the railroads promoted opportunities for American safaris Figure 2.8).

The poster child for this exploitation of wildlife was, of course, the American bison (*Bison bison*). Before white settlement, the number of bison in North America was estimated to be 40 to 50 million.[22] After the Civil War, commercial bison hunters traveled out west to exploit these animals. Bison did not react to loud noises, and an accomplished bison hunter could take down a hundred or more animals from a single herd by keeping himself concealed and using a long range rifle. After the slaughter, the hunters would skin the bison and take their tongues and hides to be shipped to eastern markets, but most of the carcasses were left to rot. As more and more people moved into the west and Amerindians were seen as an increasing menace and nuisance, hunters were commissioned simply to kill as many animals as they could as a means of starving the native. General Philip Sheridan was

FIGURE 2.8 Slaughter of buffalo encouraged by train companies. *"Slaughter of Buffalo on the Kansas Pacific Railroad" reproduced from* The Plains of the Great West, *by permission of the author, Col. R. I. Dodge. Copied from http://www.gutenberg.org/files/17748/17748-h/17748-h. htm#slaughter.*

quoted as saying, "Let them kill, skin, and sell until the buffalo is extermi-nated, as it is the only way to bring lasting peace and allow civilization to advance".[23] By 1884, only 1200 to 2000 American bison remained.[20]

Like the beaver, there was another group of animals that suffered because they provided objects that became highly desirable in the fashion trade. This group was represented by birds that had showy plumage which was incorpo-rated into women's hats as the height of fashion in the late 1800s. The pre-ferred species included egrets, pelicans, cranes, storks, ibises, and others with long feathers (Figure 2.9), but the selection did not end there. Tober (1981) quoted a source from 1886 who listed the avian species he saw repre-sented in the hats of 11 women on a horse car:

(1) heads and wings of three European starlings; (2) an entire bird (species unknown), of foreign origin; (3) seven warblers, representing four species; (4) a large tern; (5) the heads and wings of three shore-larks; (6) the wings of seven shore-larks and grass finches; (7) one-half of a gallinule; (8) a small tern; (9) a turtle-dove; (10) a vireo and a yellow-breasted chat; (11) ostrich-plumes.

Wading and sea birds were so favored that populations along the East Coast were being decimated by plume hunters. There was a bright side to this, however, in that the first wildlife refuge, Pelican Island, in Florida, was cre-ated through the efforts of President Theodore Roosevelt; this refuge was to become the first National Wildlife Refuge in the nation.

FIGURE 2.9   Bird feathers were used extensively for ladies' millinery during the 1800s.

During this time of squandering and abusing natural resources, Canada and the United States were following a similar course. The only difference was that the United States had a higher human density than Canada and an independent government, while most of Canada was comparatively sparsely populated and the country was still considered a possession of England until 1867, and then under varying degrees of English governance until 1931. A large majority of Canadians concentrated into the southernmost portion of the country eased the pressure on the northern regions, which still has a large percentage of pristine habitat. Canada also lagged a little behind the United States in its abuse of the land, because the Canadian Pacific railway (and thus the access to the west it provided) was not completed until 1885, about 16 years after the transcontinental railway in the United States.

SEEDS OF CONCERN

During the 1890s and early 1900s, the shoots of concern for natural resources were popping up. It was not that at some point everyone said "our resources are disappearing, we have to do something"; rather, the realization came over time as several resources – fish, wildlife, trees, good soil – became increasingly difficult to find. For some naturalists, such as Theodore

Roosevelt, George Bird Grinell, John Muir and others who were somewhat ahead of their time, concern began in the 1880s. For most of the populace in North America, however, concern became apparent 15 to 20 years later. By this time many species, such as white-tailed deer (*Odocoileus virginianus*), wild turkeys (*Meleagris gallapavo*), passenger pigeons, American bison, elk (*Cervus wapiti*) and beaver, were greatly diminished. Coal mines, both strip and deep underground, had scarred large parts of Pennsylvania, West Virginia, Kentucky, Alabama and Illinois. Coal centers in Canada included British Columbia, Alberta and Nova Scotia. Surface mines in particular despoiled the landscape. This mining process inverted the entire soil profile, with fertile topsoil lying at the bottom of spoil banks, covered by less fertile subsoils. Oxidation of the minerals in the exposed subsoils developed into acidic runoff that polluted streams and made the land unfit for any use. Mining companies would plunder a section of land and move on, without any concern for the environment. It was not until the 1970s that improved methods and regulations for restoration would remedy the problems caused by surface mines. We still have scarred areas that were mined before then. Today, however, the problem has not changed greatly; mining companies are removing entire mountain tops in the Adirondacks, leveling valleys and filling in streams to remove coal – and many consider this to be even worse than the early surface mining (Figure 2.10).

THE ERA OF PROTECTIONISM

In the late 1800s and early 20th century, we did not have the science for effective management of wildlife or landscapes. Instead, the concern was to

FIGURE 2.10   Moutaintop mining of today. The photo shows a mountaintop removal mining site.

preserve some of the natural resource wonders while we still had them. *Preservation* can be very important as a first step in protecting resources but, by definition, preservation is a no-interference policy. Most landscapes and wildlife populations respond better to *conservation*, which implies some management and maintenance to retain their ecological value. This period saw many firsts. In the United States:

- In 1903, the first National Wildlife Refuge, Pelican Island, Florida, was established
- In March 1909, another 52 biological reservations were protected as the seeds of the National Wildlife Refuge System
- In 1864, the first state park, Yosemite, California, was created
- In 1872, the first National Park, Yellowstone, was established
- In 1890, Yosemite came back to the federal government as the second national park
- Other early national parks included Mesa Verde NP, Colorado (1906), Sequoia NP, California (1890), Devil's Tower National Monument, Wyoming (1906), and Petrified National Forest Monument, Arizona (1906)
- Between 1891 and 1892 President Benjamin Harrison authorized the establishment of 15 forested areas encompassing 13 million acres (5.3 million ha), which eventually became part of the US Forest Service in 1905.

At about the same time in Canada:

- In 1885, the first national park, Banff in Alberta, was created
- Other early parks included Yoho, British Columbia (1886), Glacier, British Columbia (1886), St Lawrence Islands, Ontario (1904), Jasper National Park in Alberta (1907), Mount Revelstoke, British Columbia (1914), and Point Pelee, Ontario (1918)
- In 1899 the Canada Forest Service was established.

Whereas the National Parks in the United States were established primarily as wilderness areas with some tourist traffic, the initial goal of Canadian National Parks was more utilitarian, with a greater emphasis on multiple use. Actual ecological preservation as a primary objective didn't occur until later. In both the United States and Canada, the railroads actively touted the beauty and splendor of the National Parks to attract visitors and greater use of their trains.

The preservation philosophy permeated wildlife management efforts during the same era. Because the science of population ecology was not well developed, state and federal agencies decided that if game species were declining, it must be due to too many animals dying. The agencies acted on population declines in two ways. First, they established more stringent harvest regulations, including totally closed seasons to reduce take. Second, they embarked on a pogrom of predator control. Predators, they reasoned, were competitors with humans for the dwindling wildlife populations; thus, if predators were removed, it would leave more animals for sportsmen. Intensive programs to eliminate wolves (*Canis lupus*), coyotes (*Canis latrans*), cougars (*Felis concolor*), eagles, hawks, vultures, and other predators were started by many state agencies. Little thought was given to habitat management or to providing the necessary resources to wildlife populations that are now stock-in-trade in wildlife management. Virtually any method — poisoning, shooting, trapping, extermination of dens — was allowed. Federal agencies were even assigned the task of developing improved methods of predator control and preventing animal damage. The early roots of the US Fish and Wildlife Service were based on the negative effects crop predation had on agricultural economics. The National Park Service also practiced predator control on some of its holdings, which clearly was contrary to the stated mission of the Service.

During the 1920s, non-government organizations such as the Boone and Crockett Club, the Sierra Club, the New York Zoological Society, and the American Society of Mammalogists began to protest the government's policies and to criticize predator elimination in general.[24] These groups were concerned that there was inadequate scientific knowledge about the effects of predator control. Gradually, total predator removal was reduced to selective predator control by federal and state agencies.

THE DAWNING OF MODERN WILDLIFE BIOLOGY

During the 1920s, an event occurred that dramatically changed the course of wildlife conservation. At that time a young biologist, Aldo Leopold (Figure 2.11), worked for the US Forest Service in Arizona. Part of the

FIGURE 2.11   Aldo Leopold, the "Father of Modern Wildlife Management".

territory he worked in included the Kaibab Plateau. This plateau stands 1.7 miles (2.9 km) above sea level and has a surface area of about 850,000 acres (344,000 ha).[25,26] It is located in northern Arizona, and currently extends from the north rim of the Grand Canyon into the Kaibab National Forest. Amerindians occupied the plateau for thousands of years before settlers of European descent came into the area. In 1909, when the US Forest Service began a program of predator elimination, the official estimated mule deer population on the plateau was 30,000 at most, although independent estimates variously placed the number of deer from 50,000 to 100,000; subsequently the US Forest Service revised its estimates to around 100,000 animals. From 1924 to 1926, massive starvation set in and at least 60% of the deer died. Notably, cattle were also being grazed on the plateau, and there was a population of elk there as well. In his account of the situation, Leopold (1943)[27] blamed the die-offs on a policy of predator removal. He explained that the deer population exceeded its ecological carrying capacity, degraded the habitat and starved because all of the nutritious food was gone. That predators were important in regulating wildlife populations and that habitat quality could be managed became two of the main tenets of the budding science of wildlife biology. We might note that in 1970 Caughley[28,29] re-examined the existing data from the Kaibab and disagreed with Leopold's conclusions. Caughley claimed that the lack of predators had nothing to with the deer population irruption, and that balance through die-off was eventually inevitable due to plant/herbivore interactions only. Other biologists have variously defended Leopold, supported Caughley's conclusions, or arrived at some intermediate hypotheses. Whatever the cause of the crash in the deer population those many years ago, the seeds of modern wildlife management were planted in Aldo Leopold's mind on that plateau. Leopold went on to write *Game Management* (1933), the first textbook on wildlife management that took a habitat approach. He also developed the nation's first wildlife department, at the University of Wisconsin, Madison. Eventually, Leopold became known as the "Father of Modern Wildlife Management".

The next few decades witnessed two world wars, the Dust Bowl Era, and the Great Depression; these events trumped concerns about the environment, and few new things were accomplished until after the middle of the century.

POST-WAR CAPITALISM AND THE MOVE TOWARDS ENVIRONMENTALISM

World War II abutted the Great Depression and, for a while, provided the employment that both the United States and Canada needed to produce a healthier economic environment. However, the jobs and products were dedicated to the war effort, and when the war was over both countries could easily have slid back into recession if not depression. In fact, Canada was harder hit by the Great Depression than was the United States, with greater cuts in productivity, higher rates of unemployment, and no social experimentation programs like Franklin Roosevelt's New Deal. Fortunately, from an economic perspective, the rebuilding of Europe coupled with domestic programs to support the financial conditions of Canada and the United States brought a surge in an emphasis on capitalism. Unfortunately, from an environmental perspective, the push for economic growth was often made at the expense of natural resources. Unparalleled rates of mineral extraction, industrial pollution of water and air, and loss of wildlife populations leading to extinction or near extinction of hundreds of species, most of which started before the war years, continued unabated. In 1948, 20 people were asphyxiated and more than 7000 became seriously ill as the result of severe air pollution over Donora, Pennsylvania.[30] Gary, Indiana became known for its colorful sunsets of greens, yellows, reds and purples − not because of natural sunlight, but because of light refracting through a haze of airborne chemicals. The Cuyahoga River in Ohio became famous for catching fire, not once but many times, and for containing a dead zone from Akron to Cleveland − a distance of about 40 miles (67 km). In 1969, after a major fire caused over a million dollars worth of losses in boats and structures, *Time* magazine called the Cuyahoga the river that "oozes rather than flows", and where a person "does not drown but decays"[31] (Figure 2.12). Acid rain from industrial combustion of high-sulfur coal, automobile exhausts and other sources polluted streams, rivers and ponds; in some areas the pH of rain approached that of vinegar. Eastern coniferous forests, snails, trout, amphibians and other organisms were all impacted by wet and dry precipitation. During this time, surface mining continued to scar the land without any reclamation.

FIGURE 2.12 The Cuyahoga River on fire in 1952. *Credit: National Oceanic and Atmospheric Administration Ocean Service.*

Along with this economic and industrial growth, much of the world was experiencing the "baby boom". In the United States, the birth rate spiked at around 26.5 live births per 1000 women. This can be compared to 14 live births/1000 today (US Bureau of the Census). The US population rose from 132 million in 1945 to 152 million in 1950 – a net increase of 15%. Canada saw a similar growth, with 400,000 babies born each year between 1945 and 1960. This population growth added to the demand for more products and greater industry.

Although some efforts to curb the pollution began in the 1950s, it was the 1960s and 1970s that saw the rebirth of serious environmental concern in the United States and Canada. The industrial productivity of postwar United States and Canada, of course, greatly contributed to the pollution that had accumulated for decades. But it was the new affluence of Americans and Canadians that freed us from the drudgery of the Industrial Revolution. Much of the population had moved out of the major cities into suburbs, somewhat closer to nature. The work week decreased to 40 hours per week. Rachel Carson's *Silent Spring* (1962)[32] was a clear bell-ringer on the perils of the use of pesticides and other contaminants in the environment. Other books, both radical and inspirational, such as Paul Ehrlich's *Population Bomb* (1968)[33] or Aldo Leopold's *A Sand County Almanac* (1949,[34] re-released 1966), hit the bestseller lists and helped to persuade a generation that the countries – indeed the planet – were in dire trouble but worth saving.

At the time, some people considered that the prognosticators of doom were exaggerating, and indeed some were. However, people were activated by the publicity and by new environmentally based non-government agencies, which were enjoying a healthy growth. Flippin (2003)[35] presented a very readable account of these years. The first Earth Day was in April 1970. The media picked up on these concerns and eventually attracted the attention of Congress, leading to several very important Acts being passed during the late 1960s and 1970s. In 1969, the *National Environmental Policy Act* (NEPA) was passed in the United States. Among other things, this Act required that all federal projects be accompanied by environmental impact statements. During the 1970s, several other powerful bills were enacted (these laws will be discussed more fully in Chapter 3):

- In 1970, Congress enacted the first Clean Air Act
- The EPA itself was established in 1970
- The Clean Water Act was passed in 1972
- The Endangered Species Conservation Act became law in 1969, although the Supreme Court determined that it exceeded federal authority and it was replaced by the current Endangered Species Act in 1973.

Canada again was somewhat behind the United States in enacting environmentally responsible laws. Whereas *Environment Canada*, the leading federal agency for environmental concerns, was established in 1971, some of the most prominent laws were not enacted until 20 years later. The Canadian Environmental Policy Act (CEPA), 1999, corresponds in part to NEPA in requiring public input into federal projects that potentially affect the environment. It also includes some provisions for regulating water and air pollution that appear in the US's Clean Air and Clean Water Acts and amendments. The Migratory Bird Convention Act (1994) adds to the authority contained within the Migratory Bird Treaty Act (1918), of which Canada is a member. The Species at Risk Act (SARA, 2002) is comparable in purpose to the Endangered Species Act of the United States.

The environmental cause in the United States took a downturn during the Reagan administration (1981 to 1989). It is somewhat surprising that so many environmental policy Acts were passed in the 1970s under the administration of President Richard Nixon, because the Republican Party typically has had a softer stance on the environment than the Democrat Party.[35] However, it is my opinion that Reagan returned the Republican Party to its low emphasis on the environment when he appointed two administrators that were definitely not friends of the environmental movement. James Watt was Secretary of the Interior from 1981 to 1983. There is not much argument that Watt was a strong advocate of reducing federal spending and authority, and wasted little time in stripping the department of many of its regulatory and oversight functions. He pushed for greater use of Interior lands for mining, timber harvesting and oil exploration, even in National Parks, National Wildlife Refuges and Wilderness sites. He also offered a lot of federal land for sale at bargain basement prices. The other person who appeared to be an anti-environmentalist of note in the Reagan cabinet was Anne Gorsuch, who administered the EPA from 1981 to 1982. She cut the budget of the EPA by 22%, deregulated environmental oversight, and ignored violations. Congress accused her of misusing $1.6 billion from Superfund accounts, although she was never prosecuted. There was a side benefit to this anti-environmentalism, however. Non-government environmental groups such as the Sierra Club, National Wildlife Federation and World Wildlife Federation experienced great increases in membership. However, the eight years of the Reagan Administration took a lot of wind out of the sails of the environmental movement – wind that has never really returned.

Since the 1990s, the strong enthusiasm for environmental support has waned a bit. I suspect that some of this is due to the successes in the 1970s that established the infrastructure for managing these problems. We now have the tools and the agencies necessary to battle environmental issues. Also, in recent years economic concerns have grabbed the attention of most Americans and Canadians.

There is another factor, and that is, for lack of a better term, "crying wolf". Environmental zealots have repeatedly raised concern over issues that have not yet materialized. That is not to say that we should not be somewhat concerned, but the frequent raising of false alarms can reduce the credibility of all environmentalists. For example, Ehrlich's book *The Population Bomb* (1968) has so far proven to be false prophesy. Other disaster books advocating the immediate decrease of the human population have been just as wrong as Thomas Malthus when he wrote *An Essay on the Principle of Population*, in 1798. The average person soon loses interest when issues are not constantly being presented to them, but they lose trust when doomsayers are repeatedly off target.

On the other hand, it seems that we never run out of environmental problems. At this writing, the big issue is global climate change — how real is it? What are its causes? What will happen to the environment as the change continues? What will be the next environmental concern in years to come?

Study Questions

2.1 Describe the various impacts of native aborigines on the environment in North America.

2.2 What were the three European countries *most* interested in the New World? How did their objectives differ from each other?

2.3 Discuss the concepts of the Myth of Superabundance and Manifest Destiny. How did the pursuit of these ideals affect natural resources in North America?

2.4 What were the chief factors that led to the era of Protectionism? What was needed to change attitudes towards a more scientific approach to conservation and management?

2.5 How would you describe the status of natural resources in North America today? Do we have adequate protection? In what ways are we or are we not using our resources wisely? How does this compare to other parts of the world, especially developing countries?

REFERENCES

1. Santayana G. *The Life of Reason*. Amherst, NY: Prometheus Books; 1905.
2. Dillehay TD. Palaeoanthropology: tracking the first Americans. *Nature*. 2003;425:23−24.
3. Denevan WM. The pristine myth: the landscape of the Americas in 1492. In: Warren LS, ed. *American Environmental History*. Malden, MA: Blackwell Publishers; 2003:5−25.
4. Warren LS. *American Environmental History*. Malden, MA: Blackwell Publishers; 2003.
5. Cronon W. 1983. Changes in the land: Indians, colonists, and the ecology of New England. Hill and Wang, New York, NY.
6. Williams, Gerald W. Introduction to Aboriginal fire use in North America. *Fire Manage Today (USDA Forest Service)*. 2000;60:8−12.
7. Anderson RC. Evolution and origin of the central grassland of North America: climate, fire and mammalian grazers. *J Torrey Bot Soc*. 2006;133:626−647.

8. Camill P, Umbanhowar Jr CE, Teed R, et al. Late-glacial and Holocene climatic effects on fire and vegetation dynamics at the prairie–forest ecotone in south-central Minnesota. *J Ecol.* 2003;91:822–836.

9. Martin P. Prehistoric overkill: the global model. In: Martin P, Klein RG, eds. *Quaternary Extinctions: A Prehistoric Revolution.* Tucson, AZ: University of Arizona Press; 1984:354–403.

10. Grayson DK, Meltzer DJ. A requiem for North American overkill. *J Arch Sci.* 2003;30:585–593.

11. Forkey NS. *Canadians and the natural environment to the twenty-first century.* Toronto: University of Toronto Press; 2012.

12. Crosby AW. Virgin soil epidemics as a factor in the aboriginal depopulation in America. *William and Mary Quart.* 1976;33:289–299.

13. Albright H.M. Lecture. Conservation Challenge of the Sixties, Berkeley CA; 1963.

14. Tober JA. Who owns the wildlife? The political economy of conservation in nineteenth-century America. *Wildl Soc Bull.* 1981;14:459–465.

15. Taber RD, Payne NF. *Wildlife, Conservation and Human Welfare: United States and Canadian Perspective.* Malabar, FL: Kreiger Publishing Co; 2003.

16. Wilson A. *American Ornithology,* reproduced in *Borland H. 1969. Our Natural World. The Land and Wildlife of America as seen and Described by Writers Since the Country's Discovery.* Philadelphia, PA: JB Lippincott Co.; 1808–1813.

17. Carlos A.M., F.D. Lewis 2010. The economic history of the fur trade: 1670 to 1870. EH. Net Encyclopedia. <http://eh.net/encyclopedia/article/carlos.lewis.furtrade>.

18. Taylor A. *Wasty ways: stories of American Settlement. Warren. American Environmental History.* Malden, MA: Blackwell Publishing; 2003.

19. Rapp V. *New findings about old growth forests.* Portland, Oregon: Pacific Northwest Research Station, US Forest Service; <http://www.fs.fed.us/pnw/pubs/science-update-4.pdf>; 2003.

20. The Nature Conservancy. The Boreal Forest. <http://www.nature.org/ourinitiatives/regions/northamerica/canada/placesweprotect/boreal-forest.xml>; 2013.

21. Rainforest Action Network. How much old growth forest remains in the US? <http://ran.org/fileadmin/materials/grassroots/ryse/OldGrowthForests.pdf>; 2013.

22. US Fish and Wildlife Service. American Bison. <http://www.fws.gov/refuge/Neal_Smith/wildlife_and_habitat/bison.html>; 2012.

23. Old West Legends. Buffalo hunters. <http://www.legendsofamerica.com/we-buffalohunters.html>; 2003.

24. Cahalane VH. Evolution of predator control policy in the national parks. *J Wildl Manage.* 1939;3:230–237.

25. Binkley D, Moore MM, Rommie WH, Brown PM. Was Aldo Leopold right about the Kaibab deer herd? *Ecosystems.* 2006;9:227–241.

26. Leopold A. Deer irruptions. *Wisc Conserv Bull.* 1943;8:3–11.

27. Leopold A. Deer irruptions. *Aspen Bibliography.* Paper 6782. <http://digitalcommons.usu.edu/aspen_bib/6782>; 1943.

28. Caughley G. Eruption of ungulate populations, with emphasis on Himalayan Thar in New Zealand. *Ecology.* 1970;51:53–71.

29. Leopold A. *Game Management.* Madison, WI: University of Wisconsin Press; 1935.

30. Pennsylvania Department of Environmental Protection; 2005.

31. "The Cities: The Price of Optimism". *Time,* August 1; 1969.

32. Carson R. *Silent Spring*. Boston, MA: Houghton Mifflin; 1962.
33. Ehrlich RP. *The Population Bomb*. New York, NY: Ballantine Books; 1968.
34. Leopold A. *Sand County Almanac and Sketches Here and There*. New York, NY: Oxford University Press; 1949.
35. Flippin JB. Richard Nixon and the triumph of environmentalism. In: Warren LS, ed. *American Environmental History*. Malden, MA: Blackwell Publishing; 2003:272–297.

Part II

Environmental Law

Chapter 3

Historical Perspectives on the "Ownership" of Wildlife

D.W. Sparling: Natural Resources Administration. DOI: http://dx.doi.org/10.1016/B978-0-12-404647-4.00001-5

INTRODUCTION

As much as some people might not appreciate them, society requires laws and regulations to help us get along and to protect individual and group rights. This is true in all aspects of society, including natural resource management. In 1968 Garrett Hardin developed a scenario of what happens when we do not have laws to govern human activities, entitled "The Tragedy of the Commons".[1] Suppose, to paraphrase Hardin, that there is a field that is held in common by a village. "In common" means that the field either belongs to no one or is owned by the entire village, and everyone has equal access to it. Suppose further that the field is used as pasture on which the villagers can graze cattle. Let's also suppose that each villager starts with the same number of cattle. Initially, the pasture is very capable of sustaining the animals. However, it is human nature to add to one's small herd over time and thus to grow individual wealth. It doesn't really involve greed, just a natural desire to grow. However, as each villager adds a cow to his herd, gradually the number of cows in the pasture increases until the pasture can no longer sustain them all, available forage is consumed, and the system collapses. Now let us substitute the pasture with a national park, clean air, clean water, or whatever resource is owned either by no one or by everyone, and we can see how important rules and laws become to maintain the common property. As Hardin stated, "The Tragedy of the Commons".

This chapter examines the laws and decisions concerning natural resource management. We will initially provide a very short review of the major types of laws and how laws are formed. We will then focus on laws concerning wildlife in the United States because wildlife is an important natural resource and because there has been a long history of laws. Finally, we will discuss in some detail a unifying concept that distinguishes wildlife conservation in North America from that of Europe and other places.

TYPES OF LAWS AND HOW THEY ARE FORMED

I do not intend to provide a course in civics here, but a brief discussion of the legal system in the United States and Canada may help in understanding the development of laws concerning natural resources. The legal systems of both Canada and the United States are based on *Common Law* (also known as case law or precedent), *statutory law*, and *regulations* or *ordinances*. Common Law developed in England over a period of centuries, just as it has in the United States and Canada. We see it today primarily in the decisions of civil courts at the state, provincial and federal levels, and, most importantly, in the Supreme Courts; in other words, in the judicial branches of government. Common Law is sometimes called the "Law of the Land" because it is based on precedent and what is accepted as law by the general public. Common Law uses the nations' constitutions as the ultimate litmus test, although individual judges often interpret the constitutions differently at different times in history. When we say that Common Law is based on precedent, we mean that the whole history of previous decisions influences current decisions − if a judge or panel of judges decides on an issue of constitutionality, they trace back through history on related decisions and base their decisions on how that history relates to the present situation.

Statutory Law is formed by the enactment of new laws by the legislative branches of the nations, states and provinces. Although terminology may differ, the process is similar at all government levels and between the two countries. Legislators develop a Bill in the House of Representatives or Senate − in Canada, the House of Commons or Senate − debate its contents, perhaps amend it, and vote on its acceptability. If accepted, the bill is passed to the executive branch, where the President, Governor or Prime Minister either accepts it or vetoes it. If it is passed by the Executive Branch, the bill becomes an *Act*.

In the United States there is clear distinction between the Executive and Legislative Branches; no person can serve in both branches simultaneously except the Vice President, who is also the President of the United States Senate. While the Executive Branch has the power to accept a bill or veto it, it does not have the authority to create bills. In the parliamentary system in Canada, as explained further in Chapter 5, the Prime Minister is head of both the Legislative Branch and the Executive Branch, and some of the ministers of the Legislature Branch are heads of Cabinet departments. Thus, the department heads can write bills and introduce them into the Legislative Branch. However, because of the balance of power that exists among the branches of government in both nations and at the state or provincial levels, these Acts are subject to review by the Judicial Branch. Using Common Law principles, the Act may be considered for its adherence to the constitution. If declared unconstitutional, the Act becomes null and void; if declared constitutional, the Act receives a strong measure of stability.

In addition to Common and Statutory Law, the Executive Branch, including its various departments or ministries, has the authority to write *regulations, ordinances* and *rules*. At the highest level, there are *Executive Orders*. Executive Orders have the full force of law, since they are typically made in relation to certain Acts of Congress, some of which specifically delegate power to the President or Governor, or they are believed to take authority from a power granted directly to the Executive Branch by the Constitution. Sometimes Executive Orders may exceed Presidential or Ministerial powers and can be overridden by the Legislative or Judicial Branch. Also, they may be easily countermanded by successive Presidents or Prime Ministers, so they may not last as long as judicial decisions or statutory Acts. Regulations, ordinances and the like set up laws or restrictions affecting harvest, fines, and other mundane aspects of society. For example, the daily and possession limits that states set on game animal harvesting fit in this category. They, too, however, must be constitutional.

STATE AND PROVINCIAL DEVELOPMENT

The individual states have extensive powers over the regulation of wildlife harvests. Unless the species are anadramous, individual states also have wide authority over harvest of fisheries. According to the first Supreme Court decisions on wildlife (*Marten v Waddell* 1842), the principal level for maintaining the Public Trust Doctrine is the state. However, it would indeed be a daunting task to review the history of wildlife law and court decisions for each state. Therefore, I won't even attempt this.

In a somewhat similar way, Canada's government is also based on a system of federalism. Canada's principle of federalism divides the government into two constitutionally autonomous levels of government: the federal or central government, and provincial governments. Except for a few federal laws associated with international treaties, protection of federal property, and endangered species, the federal government has little to do with wildlife conservation or the management of lands in the provinces. Much of the important legislation, therefore, occurs at the provincial level. Again, individual provincial laws are beyond the scope of this book. Fortunately, for those who are interested in this topic, Donihee (2000)[2] has written an excellent province-by-province review of major events in Canadian wildlife law. In general historical development Canada paralleled the United States, but around 20 to 50 years later than the US.

DEVELOPMENT OF WILDLIFE LAW – COLONIAL TIMES AND EARLIER

In early England and most of Europe, before the discovery of the Americas, all wildlife belonged to the nobility. In fact, before the 13th century in England, all lands and all resources on these lands belonged to the monarchy. Parcels of land were bequeathed to nobles as long as they were in favor with royalty; if they fell out of favor, everything they owned under this feudal system could be taken away. Naturally, the peasants owned nothing; all that they possessed had been given to them by the noble on whose land they resided. By 1250 the nobles of England had grown tired of total fealty to the king, and they organized a protest which led to the development of the *Magna Carta*. This document was the first law to give specific freedoms and liberties to the nobility, and declared that the king was subject to English law just like anyone else. (Of course the serfs were still left out, but that was the way at the time.) Notably, the freedoms guaranteed by the Magna Carta were extended through the charters that were issued by the king. Some of these charters directly gave rights to the colonists themselves. The *Massachusetts Bay Company* charter, for example, stated the colonists would "have and enjoy all liberties and immunities of free and natural subjects." Likewise, the *Virginia Charter* of 1606 similarly declared that the colonists would have all "liberties, franchises and immunities" as if they had been born in England.[3] When the Massachusetts General Court drew up the laws for the colony, it relied on the Magna Carta as the chief embodiment of English Common Law. Thus these charters conveyed authority to make laws for the people and provided that lands and resources could actually be owned by the colonists. English Common Law, however, declared that wild animals belonged to no individual until *reduced to possession*. This phrase is derived from ancient Greek statutes, and means that until a wild animal is in the hands of a person it belongs to no one; however, once in possession, laws can be applied to the manner of capture, number caught, and the like.[4]

Eventually, however, rules needed to be established, even in early America, to prevent over-exploitation of wildlife. The colony of Connecticut placed a bounty on wolves in 1647 to protect the deer population and humans. Boston enacted the first hunting seasons for wildlife in 1698, and New York established the first warden system in 1739.[4] In the late 17th and early 18th centuries other colonies quickly followed, with closed seasons on specified wildlife. Early protection extended to deer, turkeys, heath hens (*Tympanuchus cupido*), partridge (grouse) and quail. It was a bit more difficult to set regulations on the harvest of fish because many ponds and lakes fell totally within private property and fish, unlike wild birds or mammals, do not simply get up and walk or fly somewhere else. Thus freshwater fish often were not part of the Commons, but some early restrictions were placed on riverine and estuarine species.[4] For the most part, however, enforcement of pre-colonial fish and wildlife laws was lackadaisical.[3]

CRITICAL SUPREME COURT DECISIONS

After the establishment of the United States in 1776, the federal and state governments needed to begin to fit the pieces of a sensible wildlife policy together. Wildlife law was not a high priority in the first few decades of the young country — other matters were far more pressing. Gradually, however, with the involvement of the various levels of government and the interaction of the government branches, it proceeded to hammer out wildlife law. An early guiding principle to the development of wildlife law was the *10th Amendment to the US Constitution*, which says that "The powers not delegated to the United States by the Constitution, nor prohibited by it to the States, are reserved to the States respectively, or to the people".[5] Early on, this was interpreted as meaning that the individual states had sole authority over wildlife regulations. However, the US Constitution recognizes three primary and exclusive powers of the federal government: (1) the ability to control interstate commerce; (2) the authority to make treaties and conventions with other nations; and (3) the responsibility of protecting federally owned land. Bean (1983)[6] wrote a very comprehensive summary of the development of state and federal wildlife law in the United States, on which much of the following is based.

An early issue that the government needed to decide was what to do with charters that had been given to individuals during colonial times concerning the harvest or possession of wild animals. Did these charters give individuals precedence over state and federal laws? This question was answered in the first US Supreme Court decision involving wild animals, *Marten v Waddell* 1842.[7] Waddell claimed that he owned all of the rights to oyster harvest from certain mudflats in the Raritan River of New Jersey (Figure 3.1), and that he owned both the riparian and submerged lands of the river due to a grant that had been bequeathed by King Charles II to the Duke of York in 1664. To Chief Justice Taney, the legal question here involved whether the King of England had the right to make this grant and, for our interest, whether this grant still pertained after independence. The Supreme Court declared that, due to the Magna Carta, the king did not have the authority to grant these lands to the Duke. Moreover, when the people of New Jersey took over the governance of the state they assumed the rights and authorities of the same. Thus the state, not individuals, assumed the role of successors to Parliament and the Crown. This decision laid the ground for the very important *Public Trust Doctrine* that wildlife belongs to the people in common, and that state governments have the obligation to protect these resources for the common good. This doctrine was to permanently shape wildlife conservation in North America.

The next major landmark in wildlife law supported states' rights over those of the federal government. In *Smith v Maryland* 1855,[8] Smith owned a ship for harvesting oysters with a drag or scoop; the ship was licensed through the United States, but the state of Maryland prohibited the use of drag or scoop for oyster harvest. The US Supreme Court supported states' rights over those of the federal government by ruling in favor of Maryland.

Strong support for individual state's rights was also apparent in *McCreedy v Virginia* 1876.[9] McCreedy, who was not a citizen of Virginia, wanted to plant oysters in Virginia waters for future harvests (Figure 3.2). Virginia, however, prohibited non-residents from harvesting oysters. The US Supreme Court declared that individual states owned the tidewaters bordering their lands AND the fish in these tidewaters. Therefore, states had the right to regulate planting or taking of oysters in their tidewaters even to the point of excluding citizens of other states. This decision served as a precedent for distinguishing resident and non-resident hunters or fishermen in a given state.

In a similar situation to that of *Smith v Maryland*, the US Supreme Court decided in favor of Massachusetts in *Manchester v Massachusetts* 1891[10] in ruling that Massachusetts could prohibit fishermen from using purse seines for catching menhaden, a type of fish (Figure 3.3). However, this time the decision was expressed less adamantly. Rather than declaring

FIGURE 3.1   The Raritan River in New Jersey, site of the legal battle in *Marten v Waddell* 1842, the first Supreme Court decision dealing with the harvest of wildlife and the beginnings of the Public Trust Doctrine. *Credit: Anthony Finley via Wikimedia Commons.*

that states *owned* tidal waters, the Court emphasized the 10th Amendment by saying that Massachusetts had the *right to control* use of navigable waters in the absence of any regulation by the United States. In addition, the Court ruled that because menhaden are used as bait to catch fish for

FIGURE 3.2 An oyster bed at low tide. This is similar to the focus of *McCreedy v Virginia* 1876. *Credit: South Carolina Department of Natural Resources.*

FIGURE 3.3 The Atlantic menhaden (*Brevoortia tyrannus*). This small fish has been the subject of some major US Supreme Court decisions on wildlife regulations.

human consumption, the states have the right to protect food used for common benefit.

The strongest Supreme Court decision for states' rights over wildlife came in 1896 with *Geer v Connecticut* 1896.[11] Geer appealed a conviction of violating Connecticut law by transporting waterfowl that had been legally harvested within Connecticut outside of the state for sale. Waterfowl at that time could be sold on the open market. Connecticut claimed that interstate transport of game was against their laws. The case was an important state versus federal issue because one of the federal authorities was to regulate interstate commerce, and the question was whether Connecticut's law infringed on that authority. The majority opinion of the Court declared that the state had the right "to control and regulate the common property of game" when the right was to be used "as a trust for the benefit of the people". Thus, states could regulate the conditions of how game was to be taken and what could happen to the game after harvest. The court went on to say that, given the unique authority of states in regulating harvest, it was: (1) doubtful that commerce had been established; (2) if commerce had been established it was clearly intrastate; and (3) if it was interstate commerce, the duty of the state to protect its food for the common good of the people superseded federal jurisdiction. Some have argued that this decision, taken in its extreme, strongly established the doctrine of state ownership of wildlife and rendered federal authority moot.

In 1900, the *Lacey Act* was passed by Congress. This was the first Act to support federal authority over wildlife, and it consisted of two parts. One part made it a federal crime to transport game that had been illegally harvested in one state across state lines and sell it in another state. Note that this law would not have been applicable to *Geer v Connecticut* because even though Geer sold waterfowl across state lines, his harvests were legal in the state of Connecticut. The Lacey Act was supported by the interstate commerce pillar of federal authority (see below). The other provision was the ability to prevent the importation of certain exotic species such as starlings (*Sturnus vulgaris*), English (or House) sparrows (*Passer domesticus*) and other species declared by the Secretary of Agriculture as nuisances.

In a case similar to *Smith v Maryland*, and one of the last to strongly support states' rights, the US Supreme Court decided in favor of the owners of a Florida sponge boat, the *Abbey Dodge*. In 1912 the *Abbey Dodge*[12] case began with the United States prosecuting the owners of a sponge boat for violating federal law because the boat was licensed by the United States, which forbade the use of divers to harvest sponges. The owners argued that Florida, which allowed divers, had jurisdiction over the taking of these sponges. The Court ruled that the federal government had the right to make this regulation, but only in waters outside of Florida's jurisdiction — that is, the state's rights usurped those of the federal government. This was the first statement by the Supreme Court that the state ownership doctrine actually precluded the federal government in making wildlife regulations. Attitudes and legal precedents with regards to federal jurisdiction were soon to change.

THE PILLARS OF FEDERAL LAW

Federal Right to Make Treaties

Recall from above that the constitution gives the federal government the exclusive right to form treaties with other nations, regulate interstate commerce, and protect its property and lands. A prime example of the right to make treaties involves migratory birds. In 1913 the United States Congress passed the *Migratory Bird Act*, which declared that all migratory and insectivorous birds were within the protective custody of the federal government and that none could be harmed or harvested without specific authority of the federal government. This included a variety of game birds, including ducks, geese, coots, rails and others. The Act only included the United States at this time. In two lower court cases the Act was declared unconstitutional, for it exceeded federal authority and infringed on states' rights. Subsequently, the United States entered into a treaty with Great Britain (representing Canada, which had not yet obtained complete independence), and the *Migratory Bird Treaty Act* was passed in 1918. Since this was a treaty and within federal jurisdiction, the 1918 Act appeared to meet constitutional muster. However, it wasn't long before this was actually tested. (More information on the Migratory Bird Treaty Act can be found in the next chapter.) In *Missouri v Holland* (1920),[13] Missouri sued a federal conservation officer, Holland, for being too aggressive in his enforcement of the Act. The state claimed that his zealous pursuit of violators was infringing on the right of Missouri citizens to hunt. The Supreme Court, however, decided that because the Act was a treaty, the federal government had the right to enforce it.

Federal Right to Protect Its Property

Several situations have supported and strengthened the federal right to protect its own property. For many years, the federal government had been given tacit authority to protect wildlife on its own property. For example, it forbade the harvest of any animal on the relatively new Yellowstone National Park and on the newly formed National Wildlife Refuges without any argument to the contrary. Eventually, this "right" was tested in the courts. In 1928, the US Forest Service wanted to remove deer from the Kaibab National Forest in Arizona because it was obvious that they had greatly exceeded their carrying capacity and the forests were being damaged. Arizona tried to prevent federally sponsored hunters from carrying out the order, and the Secretary of Agriculture sued Arizona to stop interference in the case of *Hunt v the United States* 1928.[14] The Supreme Court's decision in favor of the federal government was based on the premise that the power of the United States to protect its resources cannot be superseded by state statutes.

Through *Hunt v the United States*, it was acknowledged that the federal government had the right to protect its own property. But what about other situations — could the federal authority go beyond direct evidence of harm? *New Mexico State Game Commission v Udall*,[15] in 1969, responded to that question. Udall, the Secretary of the Interior, had ordered the harvest of deer from Carlsbad Caverns without regard to state regulations and strictly for research purposes, not because they were causing any specific harm to the environment. New Mexico sued the federal government because of the lack of damage. The Supreme Court ruled in favor of the federal government,

FIGURE 3.4 Wild horses and burros on BLM lands were the subject for the Wild Free-Roaming Horses and Burros Act 1971. *Credit: Bureau of Land Management.*

saying that it could determine which animals *may be detrimental* to the use of its property.

Further support for the legitimacy of the federal government to regulate wildlife on its lands came through the Wild Free-Roaming Horses and Burros Act 1971. Congress passed this Act because they deemed that wild horses and burros on western lands were historically symbolic of the Old West (Figure 3.4). Both species were introduced, so it was not a matter of protecting rare native species. A rancher who had a permit to graze livestock on federal land petitioned the government of New Mexico to remove some burros because they were competing for forage with his livestock. New Mexico complied, and began selling the animals. The feds demanded recovery of the burros, but a lower court decided that the federal Act was not constitutional because the Act did not protect land but only animals. However, in *Kleppe v New Mexico* 1976[16] the Supreme Court overturned the lower court's decision, stating that protection of federal lands was sufficient but not necessary for the federal government to regulate wildlife harvest.

Federal authority has continued to grow. In a lower court hearing, United States v Brown 1976,[17] the right of the federal government (this time the National Park Service) to regulate use of upstream waters flowing into parks was upheld to prevent damage from pollution. *Douglas v Seacoast Products* 1977 had similarities to *Manchester v Massachusetts* and the *Abbey Dodge*, but had an outcome that was the exact opposite of those earlier decisions. This case also involved menhaden and boat licensing. The state of Virginia prohibited non-residents from fishing for menhaden, but the fishing boat in this case was licensed by the United States. The owners claimed that Virginia's law violated the federal power of interstate commerce, and won. The majority decision not only upheld the supremacy of the federal government, but also echoed ancient Greek law and English Common Law by saying that no one could claim ownership of wild animals until they were *reduced to possession.*

Federal Jurisdiction over Interstate Commerce

The Lacey Act, the first federal law pertaining to wildlife based on the concept of interstate commerce, was actually supported, in part, by the *Abbey Dodge* case. While Florida could control harvest within its own borders, the decision did support the federal jurisdiction outside of the state. Today the Lacey Act is often used to prevent endangered or threatened species from coming into the country. For example, in 2013 the US government cited the Lacey Act in suing the Gibson Guitar Company for having a supply of mahogany and other rare woods which the government said was from Madagascar and therefore a violation of an international trade agreement.[18] Many other Acts have provisions to restrict interstate transportation of animals, whether intact or in part, and these Acts have been tested by even more court cases, so the right to regulate interstate commerce has been well established.

A Fourth Federal Right

While the Constitution granted the federal government certain obligations or rights, it could not foresee all aspects of the future. Certainly, the Founding Fathers would not have conceived that within 200 years many species would be experiencing severe population declines and be on the verge of extinction. In 1973, the Endangered Species Act was passed. We will discuss this powerful Act in greater detail in Chapter 4, but, in the current context, this Act gave the federal government a fourth pillar of jurisdiction: federally endangered species.

One of the first test cases for this Act was *Tennessee Valley Authority v Hill 1978.* The Tennessee Valley Authority (TVA) is a quasi-government agency designed to build and maintain dams on the Tennessee River and its tributaries to provide electricity to the rural Southeast. In 1967, the famed multi-million dollar Tellico Dam project on the Little Tennessee River was started. Progress was halted on the project after 1973 because the endangered snail darter (*Percina tanasi,* Figure 3.5A) was found upstream. The Supreme Court upheld a lower court decision that continuance of the project would exterminate an endangered species. Later, the species was found in other areas and transplanted to other streams so that the Tellico Dam project could be continued. Today, the snail darter has been downlisted to "threatened" status.

The Endangered Species Act has been court-tested many times, but one decision stands out — not only because it defended the Act but also because it supported federal rather than state jurisdiction on endangered species. This test was *Palila v Hawaii Department of Lands and Natural Forests.*[19] The

FIGURE 3.5 (A) The snail darter (*Percina tanasi*) and (B) the Tellico dam, which it almost prevented due to the Endangered Species Act. *Credit: Watershed association of the Tellico Reservoir.*

problem began centuries ago. The first inhabitants of the Hawaiian Islands brought pigs and Polynesian rats (*Rattus exulans*) with them. Some pigs became feral and began digging up the landscape, and the rats were omnivorously eating birds' eggs, insects, and other native flora and fauna. In the 1700s, the problem was exacerbated when Captain Cook "discovered" the Hawaiian chain and introduced domestic goats to serve as a source of meat when other ships came to the islands. Sometime over the course of the next couple of hundred years, European mouflon sheep (*Ovis aries*) were introduced for sportsmen. By the time the Endangered Species Act had been passed, much of the Hawaiian habitat had been damaged and scores of native birds had become endangered or extinct. One endangered species, the palila (*Loxioides bailleui*, Figure 3.6), was one of the species that had experienced declines but still existed. Palila live in dry forests on the Big Island of Hawaii, feed extensively on the seeds of the legume tree mamane

FIGURE 3.6 The palila (*Loxioides bailleui*), the subject of an important test case for federal jurisdiction of endangered species. *Credit: US Fish and Wildlife Service.*

(*Sophora chrysophylla*), and nest in naio trees (*Myoporum sandwicense*). The legume saplings are very attractive foods to both mouflon and pigs. The federal government sued the state of Hawaii to remove all mouflon and pigs from state lands because of the endangered bird. This was a test case, because it involved the federal government telling a state what it could and could not do on state land with regard to endangered species. The federal court upheld the federal position, stating that "the importance of saving such a national resource may be of such magnitude as to rise to the level of a federal property interest".

MORE RECENT DEVELOPMENTS

As explained in Chapter 1, the 1960s and 1970s were a time of heightened environmental awareness, and many federal and state Acts were passed to protect the air, water and other resources. During the 1980s and into the 1990s environmental policy became highly politicized, and efforts to take the bite out of environmental laws were highly contested.[20] During some periods agencies were very active in establishing new regulations, while in other years non-governmental organizations sued federal agencies to make them adhere to the responsibilities given to them by Congress. In *Trustees for Alaska v Hodel* 1986,[21] for example, the Trustees demanded that the Department of Interior involve full public participation in its environmental impact statement (EIS) regarding the Arctic National Wildlife Refuge. Further transparency to the public was mandated by *Conner v Buiford* 1988[22] regarding the sale of oil and gas leases in national forests. Sometimes federal agencies are sued for not doing their mandated jobs or violating laws. For example, the Environmental Protection Agency was found to have violated the Endangered Species Act in *Defenders of Wildlife v EPA* 1989.[23] The lower courts determined that the EPA was guilty of killing endangered wildlife by allowing the registration of strychnine to kill predators. *Sierra Club v Lyng* 1988[24] found that Forest Service had detrimentally impacted the habitat of the endangered red cockaded woodpecker by its timber-cutting practices in the national forests of Texas. Even the Fish and Wildlife Service was told, in *Defenders of Wildlife v Andrus* 1977,[25] that under the Endangered Species Act it must "do far more than merely avoid elimination of protected species. It must bring species back from the brink so that they may be removed from the protected class, and it must use all methods necessary to do so."[25] A very controversial and hotly contested wildlife law issue is the critical habitat clause of the Endangered Species Act (see Chapter 4). This clause has the power to prevent new and existing use of property if it is deemed essential habitat for an endangered species. Law suits are filed almost monthly by disgruntled property owners or users.

WHAT'S THE BOTTOM LINE IN THIS HISTORY?

One meaning that can be derived from this history is that the laws and attitudes towards natural resources have changed through time. In the first 150 years of this nation, states' rights were pre-eminent and Supreme Court decisions tended to support the right of states to regulate wildlife harvests. Thus they were actively following the 10th Amendment to the US Constitution. After 1900, however, the jurisdiction of the federal government gradually began to be recognized and upheld under certain conditions. We now recognize that the states have considerable authority to promulgate and enforce wildlife laws within their own borders. However, the Constitution grants the federal government three basic rights, dealing with international treaties, interstate commerce, and protecting its property. Over the years this latter right has been interpreted to mean that the federal government has supremacy over its property and can enact laws concerning these lands even if they are contrary to state laws. The Supreme Court has also defended a fourth right of the federal government: jurisdiction over federally listed threatened and endangered species. Finally, the answer to the question "who owns wildlife?" is the same as it was in 300 BC — no one owns wildlife until it is *reduced to possession*. In the next section we shall see how this has developed into what is called the North American Model and how this differs from its European roots.

THE NORTH AMERICAN MODEL OF WILDLIFE CONSERVATION

Before we describe the formal model, let us describe the conditions that it is based on. In the United Kingdom and many of other European nations, there is very little public land. As we mentioned regarding the Magna Carta earlier in this chapter, land ownership during medieval times was exclusive to the nobility, especially the monarchy. When death or poor financial management led to the availability of property, other nobles would acquire it. Later, a wealthy business class of commoners developed in western Europe and they were able to acquire land as well. As a result, almost all open land is owned by the very wealthy. On some estates, hunting is a prized activity. Managers are hired to maintain a high quality habitat, stock wildlife as necessary, and otherwise provide good hunting experiences for landowners and their associates. Some of these estates have been opened to fee hunting, in that, for a price, sportsmen can join hunting parties and participate in a wildlife harvest (Figure 3.7). The estate owner has a substantial say in which species and how many animals can be harvested, and the game managers accompany the hunting party to make sure that everyone abides by the rules. In other areas, sporting clubs, some more than a century old, lease hunting rights from a landowner, manage the property for wildlife, and hold controlled hunts for members only. Membership fees can be quite high, although they can include the use of rather extravagant facilities. Most people have little to no ability to pay the fees or dues, so hunting is largely reserved for the wealthy.

In contrast, in the United States and Canada hunting is open to virtually everyone. While there may be laws regulating how old a person must be to hunt without an adult, farm boys or girls can grab their shotgun and go rabbit or squirrel hunting on their land during open season. Hunters of bigger game can do the same, or travel to nearby public lands and try their luck at deer, moose (*Alces alces*), elk, bear and other species in their region. For more exotic hunts, sportsmen have the freedom to travel to other states or provinces and try for game not locally available. Once the required licenses or permits are purchased, there has always been ample freedom and space to participate in hunting game in both countries.

FIGURE 3.7   Chateau de Chambord – a 16th century hunting lodge built by King Francois I at Chambord, Loir-et-Cher, France. Later, it was used by King Louis XIV.

This freedom has led to the development of a concept called the *North American Model of Wildlife Conservation* (NAM), which was first formally expressed by Geist and colleagues[26] but has since permeated the philosophy of wildlife management in both the United States and Canada. The model describes a condition that was not intentionally planned by someone or conjured up by some group as an objective to attain. Rather, it describes a process of natural evolution that has occurred in both countries and which makes wildlife conservation markedly different than in Europe.[27,28] According to its proponents (for example, Prukop and Regan 2005[29]), the model is an ethical, sustainable approach to management that is citizen-based and supported by government funding and regulation. The model has its roots in the Supreme Court decision of *Marten v Waddell* (1842), which, if you recall, instituted the Public Trust Doctrine that wildlife belong to everyone and no one in particular, and that the states serve as the representatives of the people to prevent misuse of the wildlife resource. Based on the decades of decisions since *Marten v Waddell*, the NAM has developed seven guiding principles or "sisters":

1. *Wildlife as public trust resources.* Wildlife are shared in common by the people of the land.
2. *Elimination of markets for wildlife.* Up until the late 1800s, wild animals could be harvested in great numbers and shipped to major cities like Chicago or New York as meat. Market hunting was considered as one of the major detriments for wildlife populations. Regulated wildlife harvest for furs is still allowed, but wildlife can no longer be sold as a source of food on the open market. Market hunting exerted tremendous pressure on wildlife populations, which has now been alleviated. Today, sportsmen and women alone are the consumers of wildlife.
3. *Allocation of wildlife by law.* To ensure that everyone has an opportunity to participate in enjoying wildlife, through harvest, observation, or other uses, governments need to provide regulation.
4. *Wildlife can only be killed for a legitimate purpose.* These legitimate purposes include food, self protection, and protection of property.
5. *Wildlife are considered an international resource.* It is very apparent that we are not isolated from the rest of the world. International trade for pets, zoos, and international laws concerning threatened and endangered species affect wildlife conservation in all civilized nations.
6. *Science is the proper tool for discharge of wildlife policy.* According to the model, the application of science is the most effective method to manage wildlife populations and direct policy.
7. *Democracy of hunting.* Everyone should have an opportunity to hunt and enjoy wildlife due to the wide availability of public lands.

Geist and colleagues (2001)[26] have emphasized a strong role of the hunter/sportsman in guiding the evolution of the North American Model.

They reasoned that much of the funding for wildlife conservation has come through hunters and fishermen. For example, there is a federal excise tax on hunting and fishing equipment that reverts back to the states for research, management, education, and habitat acquisition. The purchase of waterfowl stamps provides funding for habitat purchases by the US Fish and Wildlife Service. Each year, sportsmen spend millions of dollars in travel, equipment purchases and other items, and those expenditures stimulate local economies. As an additional boon, supporters of the NAM point to the large number of sportsmen and women that have promoted wildlife conservation through the formation of organizations, publications, and activism. Many of the early leaders in natural resources conservation, such as Theodore Roosevelt and Gifford Pinchot, were sportsmen.

To be fair, there are critics of the American Model. Nelson et al. (2011)[30] summarized some of the arguments against using the North American Model as *the* representative model of wildlife conservation. They suggested that there are several inadequacies in the model as it now stands. For instance, they argue that the history cited as leading to the model does omit several factors, including the importance of non-consumptive users of wildlife, such as John Muir, in formulating early wildlife and natural resource policy. Critics also suggest that the developers of the model stopped their historical development early and did not include events that occurred in the mid-1900s and later as being important in shaping conservation. They also take issue with the specific elements of the model, saying that they are incomplete, vague, or otherwise inadequate.

So what can we conclude regarding this important doctrine? Both supporters and detractors have their points. The model does describe the end result of a long history of events that led to the way hunting has come about in North America. No one set out to design the system we now have. Through a natural evolution, the United States has maintained ample public lands on which to hunt and recreate. The Second Amendment of the Constitution protects the ownership of guns, which allows hunters to possess rifles for hunting. The independent spirit of the American and Canadian public certainly contributed to the freedom we enjoy today. To be sure, the situation in the United States and Canada and the attitude towards wildlife are very different than what we find in Europe. Most of Europe relies on a system that dates back many centuries, to when all the land was owned by the nobility. At that time, although wildlife was considered to be owned by no one in particular, land owners had exclusive use of the animals on their estate. Many owners took pride in the productivity of their lands, and hired gamekeepers to manage the habitat, provide caged reared birds for release, and keep records of each harvest. The system has not changed appreciably today, except that anybody who has the money can now purchase a hunt, similar to some of the private game farms we find in North America. However, the North American system is based on open hunting areas, fair chase, and equal opportunity.

Study Questions

3.1 How does Garrett Hardin's Tragedy of the Commons fit into the need for regulatory legislation on the use of natural resources?

3.2 Describe the difference between Common Law and Statutory Law. Give examples of both types of laws.

3.3 Trace the history of judiciary decisions concerning wildlife conservation in the United States. Why is the period around 1900 significant as far as trends in judicial decisions go?

3.4 What are the exclusive rights given by the US Constitution to the federal government in making laws? How do Endangered Species fit into this mix?

3.5 List and describe the seven "sisters" of the North American Model of Wildlife Conservation. How does this model differ from that found in most of Europe?

3.6 Discuss the *de facto* differences between the United States and Canada in the powers given to the provinces (states) and the federal government with regard to wildlife conservation. Which country seems to have a stronger federal involvement in wildlife regulation?

3.7 Who owns wildlife in North America?

REFERENCES

1. Hardin G. The tragedy of the commons. *Science*. 1968;162:1243–1248.
2. Donihee J. The evolution of wildlife law in Canada. CIRL Occasional Paper #9, Canadian Institute of Resources Law. University of Calgary, Calgary, Alberta; 2000.
3. Hazeltine HD. The influence of Magna Carta on American constitutional development. In: Malden HE, ed. *Magna Carta Commemoration Essays*. 1917:194.
4. Tober J. *Who Owns Wildlife?* Westport, CT: Greenwood Press; 1981.
5. US Constitution, 10th Amendment.
6. Bean MJ. *The Evolution of National Wildlife Law. Revised & Expanded Edition. Environmental Defense Fund.* New York, NY: Praeger; 1983.
7. US Federal Register (16 Pet.). 1842; 367.
8. 59 US Federal Register (18 How.). 1855; 71.
9. 94 US Federal Register 391. 1876.
10. 139 US 240. 1891.
11. 161 US 519. 1896.
12. 223 US 166. 1912.
13. 252 US 416. 1920.
14. 278 US 96. 1928.
15. 410 F.2d 1197 (C.A.N.M. 1969).
16. 426 US 539, 541. 1976.
17. 552 F. 2d 817 822 (8th Circuit Court, 1977).
18. Department of Justice. Gibson Guitar Corp. Agrees to Resolve Investigation into Lacey Act Violations. <http://www.justice.gov/opa/pr/2012/August/12-enrd-976.html>.
19. 639 F.2d 495. 1981.
20. Musgrave R. 1998. Federal Wildlife Law of the 20th Century. Center for wildlife law, East Lansing, MI.

21. 806 F.2d 1378 (9th Cir. 1986).
22. 848 F.2d 1441 (9th Cir. 1988).
23. 882 F.2d 1294 (8th Cir. 1989).
24. 694 F. Supp. 1260 (E.D.Tex. 1988), affirmed in part, vacated in part, 926 F.2d 429 (5th Cir. 1991).
25. 428 F. Supp. 167, 170 (D.D.C. 1977).
26. Geist V, Mahoney SP, Organ JF. Why hunting has defined the North American model of wildlife conservation. *Trans NA Wildl Natur Res Conf.* 2001;66:175–185.
27. Geist V, McTaggart-Cowan I. *Wildlife Conservation Policy.* Calgary, Alberta: Detselig Press; 1995.
28. Geist V, Organ JF. The public trust foundation of the North American model of wildlife conservation. *Northeast Wildl.* 2004;58:49–56.
29. Prukop J, Regan RJ. The value of the North American model of Wildlife Conservation – an IAFWA position. *Wildl Soc Bull.* 2005;33:374–377.
30. Nelson MP, Vucetich JA, Paquet PC, Bump JK. North American Model: An inadequate construct? *The Wildlife Professional.* 2011;58–60.

Chapter 4

A Closer Look at Key Environmental Laws

INTRODUCTION

There are several hundreds of Acts passed at the state, province and federal levels concerning natural resources in the United States. Hence a complete review of these laws would be huge – not to mention boring, except to a student of environmental law. In this chapter, therefore, we discuss a relatively few but important laws dealing with natural resources. We focus at the

D.W. Sparling: Natural Resources Administration. DOI: http://dx.doi.org/10.1016/B978-0-12-404647-4.00004-0

federal level in the United States. Each state has legislation that concerns hunting, fishing, mining, pollution, and all other sorts of environmental use and abuse. Much of this legislation is either similar to federal laws or is focused on specific situation unique to that state. Therefore, a review of some major federal laws can also provide some insight into state legislation. As we have mentioned before, Canadian law has given provinces equal if not superior stature to the federal government where natural resources are concerned.[1] But, again, it would require far more space than we have available to discuss laws province by province so we will confine this discussion to Canadian federal laws as well. In this chapter we start with some major laws concerning wildlife then branch off into other natural resources.

LEGISLATION DEALING WITH ENDANGERED SPECIES

The US Endangered Species Act and Forerunners

The US *Endangered Species Act* (ESA) is arguably the strongest wildlife law promulgated in the United States.[2,3] It is also one of the most controversial wildlife laws in the land.[2] It protects all species of plants and animals; redefines and expands the meaning of "take" or "harm"; defines the terms "endangered", "threatened", and "critical habitat"; requires that agencies develop plans to bring species back from the brink of extinction; is applicable to all federal agencies; enables international treaties and conventions on endangered species; AND has teeth — significant financial and prison terms for violators. That's a lot in any wildlife law! Here, we cover certain aspects of the ESA; additional information can be found in Bean (1983[4]).

There were a couple of precedents that led up to the current ESA. *The Endangered Species Preservation Act* (1966) was enacted by Congress to list declining species as endangered, and to give them a modicum of protection.[5] The Act directed the US Fish and Wildlife Service (FWS), US Department of Agriculture and Department of Defense (ostensibly because of the amount of land the agencies managed) to protect listed species and, to the agencies' ability, their habitats. The FWS was also directed to purchase land for the protection of endangered species. The Act did not really determine what protection meant, provide any penalties for violations, or go beyond the authority of these agencies. As history has shown us (see, for example, Chapter 2), if you do not know what to do to manage a species or its habitat, then at least protect it.

In 1969, the Act was amended as the *Endangered Species Conservation Act*. The amendments included a provision to recognize internationally endangered species and to prevent their importation. The revised Act led to the United State's participation in the 1973 *Convention on International Trade in Endangered Species of Wild Fauna and Flora* (CITES), and to an international trade agreement involving 80 countries. The revisions also included other vertebrates besides fish, birds, and mammals, and provided the first federal protection for amphibians and reptiles. Still, there were many weaknesses in the new Act and further provisions were needed for greater protection.

In 1973, Congress greatly modified the Endangered Species Conservation Act into the *Endangered Species Act*. This new law and its subsequent amendments (1978, 1982, 1988, 2004, 2009) placed listed species under the auspices of the FWS or, in some cases, the National Marine Fisheries Service (NMFS), and provided many safeguards not present in its earlier versions. For example, it defined *endangered* as "any species which is in danger of extinction throughout all or a significant portion of its range" (Figure 4.1A). For endangered species, prescribed infractions were broadly defined as "harassing, harming, pursuing, hunting, shooting, wounding, killing, trapping, capturing, collecting or to attempt to engage in any such conduct". The Act went on to define *threatened* as any species "which is likely to become endangered throughout all or a significant portion of its range in the foreseeable future" (Figure 4.1B).[6] For threatened species, specific safeguards are established on a species by species basis – in some cases full protection is provided, in others limited harvesting may even be allowed. *Critical habitat* was defined as "the portion of an area occupied by a listed species, the entirety of such an area, or even areas outside the currently occupied area".[2] It is common for a listed species to occupy an area considerably smaller than its historic range. Thus the historic range that still has potential for recolonization can be part of the critical habitat even though the species no longer occurs there. Penalties for violating the law can be a maximum fine of up to $50,000, imprisonment for one year, or both, for each infraction. For threatened species, violations can be fined up to $25,000 with 6 months in prison per violation. An infraction or violation can be considered

(A)   (B)

FIGURE 4.1   The Pocketbook mussel (*Lampsilis cardium*) (A) and Dudley Bluffs Bladderroot (*Physaria congesta*) (B); federally endangered and threatened species, respectively. *Credit: US Fish and Wildlife Service and US Geological Survey.*

a single individual event, so if an illegal activity resulted in the harm of a group of animals or plants the penalties could be severe. Penalties can be imposed without evidence of intent to harm. The current law also extends to all federal agencies, with some exceptions identified for the Department of Defense. A powerful component is that the ESA permits populations and subspecies to be given endangered or threatened status without requiring the same status for the entire species. For example, the cougar (*Puma concolor*) as a species is not listed, but the isolated subspecies *P. concolor coryi*, or Florida panther, which lives in the Everglades of Florida, is endangered.

The restrictions placed by the ESA have often become the basis for legal action. The section dealing with critical habitat in particular has become a source of contention, often because the concept of critical habitat is misunderstood. The designation of critical habitat affects federal activities more than it does personal business or homesteading: "Critical habitat designation has no effect on situations where a Federal agency is not involved − for example, a landowner undertaking a project on private land that involves no Federal funding or permit".[4] If a proposed activity involves federal loans, backing or permits, it could be subject to review by the FWS or NMFS, which can prohibit federal involvement if the project jeopardizes the current or prospective habitat of the listed species.

Because the ESA is a strong law, it has received pressure from both developers and landowners to weaken it, and from environmentalists to strengthen it or provide stronger enforcement. On occasion the FWS has been criticized for being slow or negligent in declaring critical habitat, based on the Act's legal provisions of "not prudent" or "not determinable".[7,8] In other cases, the declared habitat has been overturned by civil courts. These actions can lead to additional lawsuits by environmental groups,[8] resulting in a vicious circle. High legal defense costs often exceed budget allocations; consequently, this leads to shifts in agency budgets from management, listing and protection to legal defense.[8] Thus there are fewer dollars to be proactive in declaring critical habitat or other protections, and this lack of ability to act leads to further law suits. Designation of critical habitat can become a political and economic football. During the administration of George W. Bush (2001−2009), for example, the FWS officially stated that "designation of critical habitat provides little additional protection to species" − a statement that has been blatantly disproven by biological data.[8] The authors also listed several occasions when the agency declined to identify critical habitat, possibly due to legal, political or economic pressure.

I grant that the critical habitat provision can become problematic to land developers and other private businesses if the designated critical habitat is extensive. For example, the critical habitat for the California red-legged frog (*Rana draytonii*) occupies a large section of California (Figure 4.2). If that entire area were to be considered an inviolate refuge for the frog, the cost in lost income and property would be astronomical. Major court battles

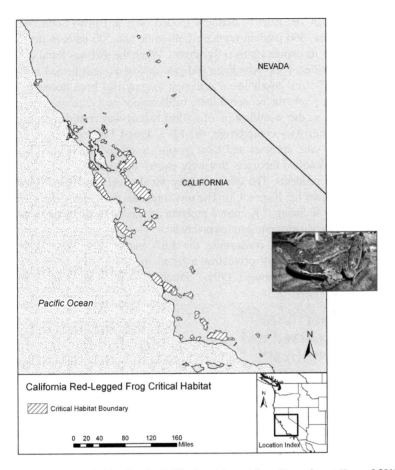

FIGURE 4.2 Critical habitat for the California red-legged frog (*Rana draytonii*) as of 2012. The designated areas encompass 1.6 million acres. Inset: the California red-legged frog. *Credit: US Fish and Wildlife Service.*

occurred when the FWS first designated 450,000 acres (182,000 ha) of California as critical habitat in 2006, but these were nothing compared to the rancor when the FWS increased that to 1,600,000 acres (647,500 ha) in 2010 (Figure 4.2). The FWS won the court battles, but conceded that the primary focus of conservation would be limited to a much smaller area.[9,10] By the way, the subject of this controversy is also the focus of Mark Twain's 1865 *The Celebrated Jumping Frog of Calaveras County*!

Another case of critical habitat issues occurred in the 1990s. The northern spotted owl (*Strix occidentalis caurina*) lives in the old-growth forests of the Pacific Northwest. Old growth here is defined by a predominance of trees at least 150 years old.[11] Largely through cutting of old-growth forests and consequent habitat loss, the species has declined over the past 100 years. As of 2012, there were fewer than 100 pairs of northern spotted owls in British Columbia, 1200 pairs in Oregon, 560 pairs in northern California, and 500 pairs in the state of Washington.[11] Its current status is threatened. When the owl was listed in 1990 it became symbolic of the remaining old-growth forests, and heated arguments arose from different stakeholders. Although logging had been declining in the area for many years, the economic base of the region still depended on logging and sawmills, so the identification of critical habitat was considered a threat to that way of life. As a compromise, the FWS, Forest Service, Bureau of Land Management, state agencies and local groups developed the Northwest Forest Plan, which identified the tracts that were closed to logging and provided assistance to local residents. The critical habitat for the owl consists of 9.6 million acres (3.9 million ha; Figure 4.3). The owl continues to decline, but at a slower rate than prior to listing.[12] A current problem for the owl is encroachment of the barred owl (*Strix varia*) into spotted owl habitat.[13]

Other controversies concerning the ESA include whether it has been effective, and the cost of protecting a listed species. Critics of the law (for example, Mann and Plummer 1995,[14] Pombo 2004[15]) often point to the few

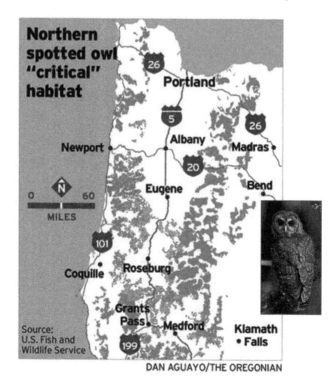

FIGURE 4.3   The designated critical habitat for the northern spotted owl (*Strix occidentalis caurina*), occupying 9.6 million acres in the old-growth forests of the Pacific Northwest. Inset: the northern spotted owl. *Credit: US Fish and Wildlife Service.*

species that have actually been delisted and to the cost per species in conserving endangered species. The ESA lists almost 1600 taxa as endangered or threatened (Table 4.1). "Taxa" in this case can be species or subspecies. There are about 1300 discrete species in these listings[3,16]; of these, 7 listed species have become extinct while the rest (99%) are still in existence. If these species had not been listed, it is predicted that up to 172 of them would have become extinct.[16] Critics also point out that only 14 species have been delisted, but delisting is a complex process that can take many generations of population growth, and insufficient time may have elapsed since the passing of the Act for a substantial number of species to have improved sufficiently to be delisted.[17] However, the number of species that have improved in their status from endangered to threatened (20) exceeds the number that have gone the other way (7) by nearly 3 to 1 (Schwartz 2008).[3]

The cost per species is extremely variable. Schwartz (2008)[3] estimated that in 2004 the top 100 listed taxa (around 10% of those listed) received 89% of the recovery funding. Those 478 species on the bottom of the spending list received less than $5000 each.[18] The imbalance continues. In 2011, the top 10 taxa received $444.8 million (35% of total state and federal expenditures). Many of the top 10 taxa were different populations of the same species; we do not reach 10 distinct species in order of expenditures until the 19th entry and a

TABLE 4.1 Number of Native Endangered or Threatened Taxa by Major Classification Listed by the US Fish and Wildlife Service (Numbers are Subject to Change through Time).

Group	Endangered	Threatened	Total
Lichen	2	0	2
Ferns	28	2	30
Conifers	2	1	3
Flowering Plants	781	160	941
Coral	0	2	2
Clams	72	11	83
Snails	27	13	40
Insects	55	13	68
Aráchnids	12	0	12
Crustaceans	19	3	22
Fish	81	68	149
Amphibians	16	10	26
Reptiles	13	24	37
Birds	77	15	92
Mammals	69	18	87

Source: http://www.fws.gov/ENDANGERED/species/us-species.html.

sum of \$611.6 million. The top two species were Chinook salmon (*Oncorhynchus tshawytscha*) and steelheads (*O. mykiss*), with a combined expenditure of \$484.4 million. In contrast, 43 species had \$1000 or less spent on them during 2011.[19] The reasons for this discrepancy in expenditures are complex. Since two species of fish have garnered a huge proportion of the funding, biology and ecological importance are not the only factors in making decisions. Certainly, politics and "likeability" or notoriety are involved to some extent. Congressmen may have their favorite organisms and push for funding, while some species are more charismatic than others.

Another argument raised by anti-environmentalists is the total amount of money spent by the Endangered Species Program. They argue that the money could be spent on other needs, such as anti-poverty programs or national defense. In the 10 years from 2002 through 2011, the federal government spent \$11.8 billion on endangered species while states spent \$60.8 million, giving a total of \$12.4 billion, or an average of \$1.24 billion per year (Figure 4.4) — a lot of money, no matter how you look at it. However, the entire budget for the Department of Interior is less than 4% of the total discretionary budget for the federal government. Some would argue that it is comparing apples and oranges, but the seven-month Desert Storm war cost approximately \$66 billion. Ultimately, how much is an endangered species worth?

The answer to that question is not really answerable — at least, there is no answer that would satisfy all Americans. Shogren et al. (1999)[20] pointed out that opinions on that question range from "very little" to that "it is not even a valid question, because the preservation of endangered species and biodiversity is of such moral importance that economics should not even enter into the discussion" (for example, Roughgarden 1995[21]). Holmes Ralston III, an environmental ethicist, posed the rhetorical question: "You wouldn't let the Ethiopians starve to save some butterfly, would you?" (Ralston 2003, p. 451[22]). He answered his own

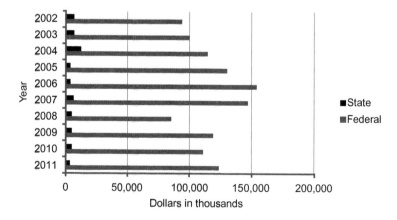

FIGURE 4.4 Annual federal and state spending on endangered and threatened species, 2002–2011. *Data from US FWS.*

question by stating that humans do make similar decisions all of the time. Only a certain fraction of the annual US budget goes towards humanitarian aid. Defense spending, research on non-human related issues, money spent on arts, music, or other areas, is money *not* spent on feeding the hungry. Therefore, if we did not spend money on conserving endangered species it is very unlikely that the savings would be spent on feeding starving humans. It might seem harsh, but it is reality. Shogren and colleagues[20] argued that economics must be considered with endangered species conservation if for no other reason than to make sure that the expenditures are being used in the most efficacious manner possible. That brings us back to asking if it is reasonable to spend one-third of the annual endangered species support on two species of fish. The issue of endangered species economics is a complex one with many aspects and is largely beyond the space we can allocate to it in this book, but it is definitely the stuff of wildlife policy.

We might better ask, what affects the success of ESA protection? Martin et al. (2005)[23] addressed this question and identified several factors affecting whether a species' status improved over time:

1. *Overall, a little more than 50% of all listed species are either remaining stable or declining.* This, however, varies across taxa, with birds faring appreciably better than other groups of plants and animals.
2. *The longer a species is listed, the higher are its chances of improving.* This could suggest that the ESA protection works cumulatively over time. It could also mean that early efforts to restore a species were more effective than later ones. In a separate analysis using different methods, Male and Bean (2005)[17] agreed with this finding.
3. *Critical habitat improves the recovery of listed species.* As mentioned above, the Department of the Interior labeled critical habitat redundant and unnecessary, stating that harm and damage to a species is sufficient protection. While this declaration slowed the declaration of critical habitat for many years, biological evidence has supported the importance of identifying critical habitat and the pace has picked up. Martin et al. (2005)[23] found that species that had critical habitat identified for two or more years appeared to have a greater probability of improving than those that did not. In contrast, Male and Bean (2005)[17] thought that the effect of critical habitat on predicting recovery was confounded by other factors and was not, by itself, very important in explaining recovery. Since this is a very controversial aspect of the ESA, further study on the advantages of critical habitat certainly seems warranted.
4. *Dedicated recovery plans based on a single species help species recovery.* Multi-species plans were less effective or not effective at all.
5. *The ESA is equally effective in improving the status of both plants and animals, but effectiveness within a biological Kingdom varied considerably.* No doubt the unequal distribution of funding discussed above is an essential factor in this finding.

6. *Endangered species show a lower recovery rate than threatened species.* This makes sense in that endangered species by definition are in more serious condition than threatened species and are likely to take longer to recover.

Overall, while the ESA has its many critics, it is making inroads into protecting rare and declining species. Can it be improved? It can certainly be improved, but not without additional funding.

Other United States Endangered Species Laws

In 1972 Congress passed the *Marine Mammal Protection Act*, which "prohibits, with certain exceptions, the 'take' of marine mammals in US waters and by US citizens on the high seas, and the importation of marine mammals and marine mammal products into the US".[24] Substantial amendments were made to this Act in 1994. The Department of Commerce, through the National Oceanic and Atmospheric Administration's National Marine Fisheries Service (NMFS), has the primary responsibility for conserving marine mammals, defined primarily as cetaceans (whales and dolphins) and pinnipeds (seals, sea lions). Several of these species are endangered and threatened, but the Marine Mammal Protection Act has an additional classification of "depleted", which refers to populations that are below optimum sustainable population levels. The Act gives the NMFS authority to develop conservation plans, form cooperative agreements with states and aboriginal groups, conduct research, organize a health and stranding response program, regulate the conditions of captive marine mammals, and work with the FWS in CITES issues. While the NMFS manages most marine animals, authority over walrus (*Odobenus rosmarus*), polar bears (*Ursus maritimus*), sea others (*Enhyrda lutris*), marine otters (*Lontra feline*), dugong (*Dugong dugong*) and manatees (*Trichechus* sp.) falls under the FWS.[25] The Marine Mammal Act provides for a fine of up to $10,000 for each violation.

The *Wild Bird Conservation Act* (1993) regulates the importation of rare and endangered birds into the United States. It focuses on species that are listed under CITES and native to other countries, to restrict their importation into the United States. However, there are several exclusions to its coverage, including domestic species; sport-hunted birds; dead scientific specimens; and birds in the families of Phasianidae (pheasants and quail), Numididae (guineafowl), Cracidae (guans and curassows), Meleagrididae (turkeys), Megapodidae (megapodes), Anatidae (ducks, swans and geese), Struthionidae (ostrich), Rheidae (rheas), Dromaiinae (emus), and Gruidae (cranes), many of which are hunted. Permits may be issued allowing importation of other avian species if they are for scientific use, zoological specimens or designated for the pet trade.

ENDANGERED SPECIES IN CANADA

The heart of federal endangered species management in Canada is a three-part strategy consisting of the *Species at Risk Act* (SARA), the *Accord for the Protection of Species at Risk*, and the *Habitat Stewardship for Species at Risk* program. The three programs unite to form a cohesive federal and provincial focus on preventing species extinctions or extirpations in the nation.

SARA is the core of this strategy. It has been called "one of the most far-reaching pieces of federal environmental legislation ever seen in Canada".[26] Through SARA the *Committee on the Status of Endangered Wildlife in Canada* (COSEWIC), established in 1977,[27] became the chief advisory board for the status of wildlife species in Canada. COSEWIC is an independent board made up of biologists and scientists, not politicians, so recommendations are made on scientific principles. The board makes annual recommendations to the Minister of the Environment on the status of species within Canada. On receipt of a species assessment by COSEWIC, the federal government can accept the assessment of COSEWIC and add the species to the legal list, decide not to add the species to the list, or refer the assessment back to COSEWIC for further consideration. The Minister makes the final determinations on status. Several important definitions were established in COSEWIC, concerning endangered wildlife species.

- *Wildlife Species*: "A species, subspecies, variety or geographically or genetically distinct population of animal, plant or other organisms other than a bacterium or virus that is wild by nature and is either native to Canada or has extended its range into Canada without human intervention and has been in Canada for at least 50 years". Organisms that fit this definition can be considered further under SARA. While the initial passage of COSEWIC only included vertebrates and vascular plants, subsequent amendments have added mollusks, some arthropods, mosses and lichens. As you may note below, the status categories for these wildlife species are a bit more complex than those in the United States ESA.
- *Extinct*: "A wildlife species that no longer exists."
- *Extirpated*: "A wildlife species that no longer exists in the wild in Canada but exists elsewhere."
- *Endangered*: "A wildlife species facing eminent extirpation or extinction."
- *Threat*ened: "A wildlife species that is likely to become endangered if nothing is done to reverse the factors leading to its extirpation or extinction."
- *Special Concern*: "A wildlife species that may become threatened or endangered because of a combination of biological characteristics and identified threats."

- *Data Deficient*: "Available information on the species is insufficient to make a determination."
- *Not at Risk*: "A wildlife species that has been evaluated and found to be not at risk of extinction given the current circumstances."

Table 4.2 provides the number of listed taxa as of 2012. Because separate populations can be individually listed, the actual numbers of distinct species that are listed are somewhat fewer than the totals in the table.

The Species at Risk Act has many important provisions, making it a powerful tool to protect wildlife species. Like the ESA, SARA can designate an entire species, or recognizable segments of a species. The species or particular subspecies can be listed in one of five categories. Penalties for violations can be as high as C$1 million and up to 5 years' imprisonment for each violation. The Act also has provisions for defining and protecting critical habitat.

The *Accord for the Protection of Species at Risk*, made in 1996, is an agreement made by federal, provincial and territorial governments to develop an endangered species strategy.[28] Among other things, the signatories of the Accord agreed to work across provincial lines, to follow federal laws concerning the conservation of these species, and to recognize COSEWIC as the chief agency for making recommendations on the status of wildlife species.

The *Habitat Stewardship for Species at Risk* is a partnership-based initiative sponsored by the Government of Canada to help Canadians protect species and their habitats. The Program is administered by Environment Canada, and is managed cooperatively with Parks Canada and Fisheries and

TABLE 4.2 Number of Listed Taxa and their Status as Assigned by Canada's Species at Risk Act in 2012 (Numbers are Subject to Change through Time)

Group	Extirpated	Endangered	Threatened	Special Concern	Data Deficient
Lichens	0	4	2	4	3
Mosses	1	7	3	4	1
Vascular Plants	3	95	48	29	4
Arthropods	3	23	6	5	3
Molluscs	2	13	1	5	5
Fish	3	26	12	25	28
Amphibians	1	8	5	7	0
Reptiles	4	15	12	7	2
Birds	2	28	23	21	2
Mammals	4	17	14	18	8

Source: http://www.cosewic.gc.ca.

Oceans Canada. Essentially, it provides the funding for many conservation projects developed for listed species. At its inception in 2000, the federal government promised C$45 million to be spent in the first five years to support the provisions of SARA. As of 2012, it was allocating C$9–13 million each year.[29]

The *Wild Animal and Plant Protection and Regulation of International and Interprovincial Trade Act* (WAPPRIITA) 1992 is Canada's law to follow the provisions of CITES (see below). It covers all three Appendices of CITES, the transport of species of concern within Canada, and the importation of species that could lead to the endangerment of Canadian species: "The Act forbids the import, export and interprovincial transportation of these species, unless the specimens are accompanied by the appropriate documents (licenses, permits). The Act applies to the plant or animal, alive or dead, its parts, and any derived products".[30] The Act can impose fines of up to C$25,000 on individuals and C$50,000 on corporations or groups.[31]

Endangered species legislation did not come easy to Canada.[26] This is a story of policy, and differing stakeholder views. It started nearly 20 years after the US Endangered Species Act was passed, when Canada signed the Convention on Biological Diversity at the Rio Earth Summit in 1992. Before the Summit the only group concerned in Canada with protecting endangered species was COSEWIC, which had been formed in 1977 and had created the Recovery of Nationally Endangered Wildlife program in 1988 to develop recovery plans. Neither COSEWIC nor the Recovery program had paid staffers, regular sources of income, or any real power beyond making recommendations. Consequently, there was not much progress made in the first years after their establishment. After the Rio Earth Summit, the Parliament's Environment Committee held hearings on how to best comply with the convention. A few key NGOs, including the Sierra Legal Defense Fund (SLDF), took the position that the convention required Canada to pass legislation to protect threatened wildlife. Environment Canada (EC), the chief federal agency for the environment, disagreed, saying that legislation was not needed; current laws were sufficient.

The Environment Committee, surprisingly, agreed with the NGOs rather than the EC and recommended that the federal and provincial governments "take immediate steps to develop an integrated legislative approach to the protection of endangered species".[32] The NGOs were delighted, but their excitement was premature. Discussions with officials from EC soon revealed that the government was reluctant to get involved with endangered species legislation. After being stonewalled, the NGOs formed the Canadian Endangered Species Coalition (CESC) and, with funding from several foundations, took the issue to the people. Public support for legislation was strong, and within a year the CESC had won a commitment from the federal government for endangered species legislation; that was in 1995, and it would be eight more years before legislation would be enacted.

CESC did not give up, and continued to generate public support. To make a long story shorter, the government continued to resist change. For one thing, they wanted cabinet members, not scientists, to decide which species should be listed. For another, the government wanted endangered species legislation to cover only federal lands, so as not to impose restrictions on private landowners. They also wanted to reduce federal involvement to a bare minimum, allowing each province to make its own rules. The scientific community was vociferous in joining the growing public sentiment for real legislation. Even the US Secretary of the Interior, Bruce Babbitt, wrote to Canada's Minister of the Environment, Sergio Marchi, to encourage him to stand up for cross-border legislation. Several other attempts by the federal government to enact insipid legislation engendered public resentment. Finally, after eight years of struggles, Parliament signed the current Species at Risk Act.[26] I have shortened the story considerably, but the gist of it is that sometimes government stands in the way of environmental progress. At such times the public, often spurred by NGOs, must stand up and demand laws that are soundly based in wise stewardship. Although this took place in Canada, similar sequences have occurred in the United States.

Endangered Species in the States and Provinces

In addition to federally listed species, most states and all provinces and territories have their own lists of endangered and threatened species. While they are required by law to recognize federally listed species, they may have their own lists for several reasons. The most common situation is for a state or province to be on the margin of a species' distribution. For example, the spotted dusky salamander (*Desmognathus conanti*) is comparatively common in the southern states of Alabama, Mississippi and Georgia, but its northernmost range just enters the southern tip of Illinois (Figure 4.5). The species is not even threatened in the southern states, but is considered endangered in Illinois. Many species that are common in the United States are listed in Canada for distributional reasons. For instance, the northern bobwhite (*Colinus virginianus*) is a commonly hunted species throughout much of eastern United States (Figure 4.6), but its distribution touches southern Ontario and hence it is considered endangered in Canada. Another reason a species may be on a state or provincial list is changes in habitat. Much of the native tallgrass prairie of the Midwest has disappeared to the plow, and species dependent on large tracks of grass may be common only in zoos.

INTERNATIONAL CONCERN FOR ENDANGERED SPECIES

There are two international elements that are extremely important to endangered species conservation. One is *The Convention on International Trade in Endangered Species of Wild Fauna and Flora* (CITES). This is an

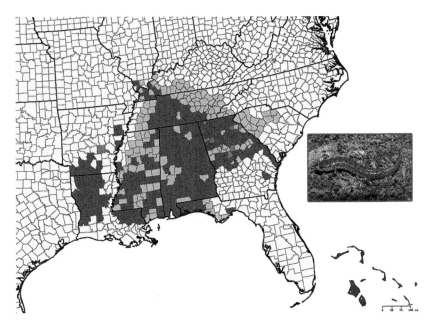

FIGURE 4.5 Distribution of the spotted dusky salamander (*Desmognathus conanti*, inset) is widespread in the Southeast but just touches southern Illinois, where it is considered to be an endangered species. *Credit: Courtesy of US Geological Survey, Photographer Brad Glorioso.*

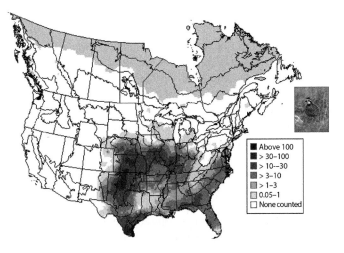

FIGURE 4.6 Distribution of northern bobwhite (*Colinus virginianus*, inset). The species is hunted in much of the southeastern United States but barely touches southern Ontario, where it is considered an endangered species. *Credit: US Geological Survey.*

international agreement established in 1973 among 177 nations to restrict the trade in plants and animals so that trade does not threaten their continued existence. "Annually, international wildlife trade is estimated to be worth billions of dollars and to include hundreds of millions of plant and animal specimens. The trade is diverse, ranging from live animals and plants to a vast array of wildlife products derived from them, including food products, exotic leather goods, wooden musical instruments, timber, tourist curios and medicines".[33] The agreement covers more than 30,000 species of plants and animals. CITES lacks the ability to impose fines or other penalties other than international pressure on member countries that violate or appear to violate its provisions. However, that pressure is often sufficient to guide a nation into compliance.

Listed species fall into one of three categories or Appendices. Appendix I contains the rarest species – those that are internationally recognized as being threatened with extinction. Trade of these animals is allowed only under very stringent and exceptional conditions. Permits are issued allowing importation only if the organism will not be used for commercial purposes, and if the specific nature of the trade agreement will not harm the survival of the species. For example, a well-respected zoo may request specimens of some animal species to enter its breeding program. Appendix II contains species that are not threatened by extinction but which require protection from harmful practices. An export permit will be allowed if the organism was taken legally and if exportation will not put the species at greater risk. Appendix III includes species that are not internationally threatened but where at least one member nation has requested a restriction on trade. Exportation is allowed with a valid permit from the native country.

The other international element concerned with endangered species is the *International Union for the Conservation of Nature* (IUCN), which was founded in 1948 as the first world environmental organization. Today it is composed of 1200 member organizations, of which 17% are governments and 83% are non-governmental.[34] The IUCN is a non-profit, nongovernmental organization whose mission is "to influence, encourage and assist societies throughout the world to conserve the integrity and diversity of nature and to ensure that any use of natural resources is equitable and ecologically sustainable". It is probably best known as the organization that publishes the *Red List of Threatened Species.* This database, which is updated annually, identifies some 66,000 species of plants and animals into the categories of extinct, extinct in the wild, critically endangered, endangered, threatened, vulnerable, near threatened, least concern, data deficient, and not examined (Table 4.3). The IUCN does not have any regulatory authority, but its Red List is used extensively by agencies, NGOs and scientists. While the IUCN may be best recognized for its list of species, it also disseminates information on the status of thousands of species, convenes meetings, builds partnerships, advises other conservation organizations, and develops standards for conservation action.

TABLE 4.3 Number of Species and their Status According to the International Union for the Conservation of Nature in 2012 (Total Included May Not Equal the Number of Species in that Taxon)

	Extinct	Extinct In Wild	Critically Endangered	Endangered	Vulnerable	Not Threatened	Data Deficient	Least Concern	Total Included
Algae	0	0	6	2	1	0	14	11	31
Fungi	0	0	4	2	1	0	9	0	15
Mosses and Liverworts	3	0	31	33	27	6	13	21	134
Ferns	2	0	35	42	64	15	50	58	265
Conifers	0	0	24	63	95	69	32	409	705
Flowering Plants	84	27	2300	4536	1480	362	1028	3097	14178
Other Inverts	2	0	2	27	214	182	157	297	891
Molluscs	310	14	549	480	828	500	1703	1793	6183
Arthropods	71	2	239	363	858	273	1817	2833	6468
Fish	60	7	415	494	1150	443	2618	5393	10591
Amphibians	34	2	509	767	657	389	1624	2392	6374
Reptiles	21	1	144	296	367	260	676	1988	3755
Birds	130	4	197	389	727	880	60	7677	10064
Mammals	77	2	196	446	497	325	834	3124	5501

Source: www.iucnredlist.org.

OTHER MAJOR FEDERAL WILDLIFE LAWS

In the United States there are a few other federal laws that should be discussed. Certainly there are many more laws that have been passed that concern wildlife, but these have had considerable impact on conservation.

The *Lacey Act* of 1900 was introduced in Chapter 2 as the first federal legislation passed to assist wildlife. It has two major components. The first is to prohibit transport of wildlife that was illegally harvested in one state or Amerindian nation and shipped across state lines. The original definition of "wildlife" was very limited, but subsequently came to include any part of a wild animal or any product made from an illegally harvested animal. In addition, the Act prevents interstate or foreign commerce involving any fish, wildlife or plants taken, possessed or sold in violation of state or foreign law. The Endangered Species Conservation Act amended the Lacey Act in 1966 to include all of the species listed by CITES, both plants and animals.[35] In 2008, another major amendment broadened the types of plants and plant products covered by the law. Now the law covers timber and products made from illegally harvested trees. This is another tool that can be used to assist the conservation of endangered species.[35] Penalties for violating the law can be several. The Criminal Fines Improvement Act of 1987 increased the fines for misdemeanors under the Act to a maximum of $100,000 for individuals and $200,000 for organizations. Maximum fines for felonies increased to $250,000 for individuals and $500,000 for organizations.[35]

The *Migratory Bird Treaty Act* 1918 (MBTA) is a very powerful law initially developed through a series of Conventions with several different nations. Based on international law, there is no legal difference between a treaty and a convention. In practice, however, treaties tend to be signed by a limited number of nations whereas conventions are more open to multiple signees. The first convention leading to the MBTA was between the United States and Canada in 1916. Great Britain was the co-signee for Canada, because Canada was not entirely independent at that time. This Convention established three categories of migratory birds: (1) those that could be harvested; (2) insectivorous birds; and (3) other non-game species. It also set a maximum open season for game species, and closed seasons for the other two groups. Arguably, the game birds of greatest interest are waterfowl – ducks and geese. However, the list of birds protected by the Act is much more extensive, and includes coots, rails, gallinules, doves and others. Insectivorous birds are of particular significance because they are major consumers of crop depredators and disease vectors. Subsequent conventions were developed between the United States and Mexico (1936), Japan (1972), and the Soviet Union (1976). Each had specific provisions, such as provisions for harvest by aboriginal groups, special protection for endangered species, and differences in the detail of recognizing listed taxa.[36]

The original Convention was formalized into the *Migratory Bird Treaty Act* in 1918. Important provisions of the Act include the following.

1. It is illegal to "pursue, hunt, take, capture, kill, attempt to take, capture or kill, possess, offer for sale, sell, offer to purchase, purchase, deliver for shipment, ship, cause to be shipped, deliver for transportation, transport, cause to be transported, carry, or cause to be carried by any means whatever, receive for shipment, transportation or carriage, or export, at any time, or in any manner, any migratory bird, included in the terms of this Convention ... for the protection of migratory birds ... or any part, nest, or egg of any such bird."[37] How many young boys or girls have been violators of this Act by "innocently" shooting a migratory bird with a BB gun, air rifle or slingshot?

2. Individual states can establish their own laws concerning migratory birds so long as they do not interfere with or weaken federal regulations.

3. There are exceptions for take by Amerindians and Eskimos, and through scientific permits.

4. Penalties for misdemeanor infractions could include a fine of not more than $500 or imprisonment of not more than six months. Activities aimed at selling migratory birds in violation of this law are a felony, and subject to fines of not more than $2000 and imprisonment not to exceed two years. Guilty offenses would constitute a felony. Equipment used for sale or purchases is authorized to be seized and held, by the Secretary of the Interior, pending prosecution, and upon conviction be treated as a penalty. Hunting over bait can result in fines of up to $100,000 for individuals and $200,000 for organizations, imprisonment for not more than one year, or both.[38]

Two Acts Not Based on Ecological Need

The Bald Eagle (*Haliaeetus leucocephalus*) was on the Endangered Species List from 1973 to 2007. Major problems leading to endangered status were eggshell thinning from the pesticide DDT,[39–41] and shooting. When it was delisted, the bald eagle was still under very strict protection due to two other Acts. One of those is the Migratory Bird Treaty Act, discussed above; the other is the *Bald Eagle Protection Act* (1940). The Act prohibits the take, possession, sale, purchase or barter or offer to sell, purchase or barter, transport, export or import of any bald eagle, alive or dead, including any part, nest or egg, unless allowed by permit.[42] "Take" includes pursue, shoot, shoot at, poison, wound, kill, capture, trap, collect, molest or disturb. In 1962 the Act was amended to include the Golden Eagle (*Aquila chrysaetos*), in part because it is very difficult to distinguish a juvenile bald eagle from a golden eagle in the field, and in part because the golden eagle deserved protection in its own right. An important precedent was set by this law, because the bald eagle was initially given protection not because it was dwindling in numbers or because of its ecological significance, but because it was the nation's symbol.

Another Act actually protects species that some would say contribute to the destruction of a fragile habitat. We discussed the *Wild Free-Ranging Horses and Burro Act* (1971) back in Chapter 2 because some environmentalists want to get rid of the Act as it offers protection to what they believe are introduced, noxious species. Others, especially historians and animal lovers, claim that the horses and burros deserve protection, for they are a vestige of when our country depended heavily on the animals for transportation and hauling. The Act states that "Congress finds and declares that wild free-roaming horses and burros are living symbols of the historic and pioneer spirit of the West; that they contribute to the diversity of life forms within the Nation and enrich the lives of the American people; and that these horses and burros are fast disappearing from the American scene".[43] The Secretaries of the Interior and Agriculture, through the Bureau of Land Management and US Forest Service, respectively, were ordered to manage the herds of feral burros and horses. From time to time the biologists of the agencies conduct a population count, and if the animals are exceeding their carrying capacity the agencies are authorized to conduct a sale of "surplus" stock. In the past this often meant that the animals would be sold to anyone, but the Bureau has changed its policy to make every effort to place the surplus in good homes or ranches.[44] As with the Bald Eagle Protection Act, this protection was established to protect a symbol of American life.

OTHER FEDERAL WILDLIFE-RELATED LAWS IN CANADA

The Migratory Bird Convention Act 1917 was the Canadian version of the United States Migratory Bird Convention Act with Canada. It was also the Canadian analogy to the Migratory Bird Treaty Act of 1918 until 1994, when it was substantially revamped. The revised law contains many provisions that the old one did not. It provides regulations on many aspects of migratory birds, including[45]:

- sale, gift or purchase
- shipment
- aviculture
- taxidermy
- activities involving birds causing damage or danger (e.g., agriculture)
- activities involving overabundant species
- activities at airports
- activities for scientific research purposes
- collection, possession, sale or trade of eiderdown
- import of migratory bird species that are not indigenous to Canada.

While it allows hunting of some migratory species, the law sets permit requirements, possession limits and seasons for game birds. One of the key provisions of the newer law is that it establishes migratory bird sanctuaries, which, along with the National Wildlife Areas established by the Canadian Wildlife Act, provide protected habitat for birds and other species. These sanctuaries and wildlife areas are scattered across Canada, but are most numerous along the eastern seaboard and the prairies of Alberta, Saskatchewan and Manitoba (Figure 4.7).

The *Canadian Wildlife Act 1985* was enacted to create and manage National Wildlife Areas for the protection of wild animals. Research, interpretation and conservation are promoted in these areas, but all forms of harvest or take of both plants and animals are prevented without proper permits. Some activities, such as nature viewing, hiking, canoeing, and wildlife photography, are allowed. The Act gives special preference to migratory species and listed species.

LAWS THAT FOCUS ON HABITAT

There are numerous laws both in the United States and Canada that are designed to protect our environment from contamination through pollution.

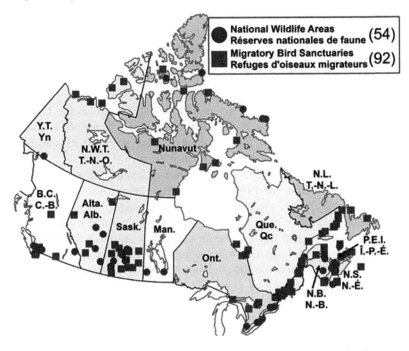

FIGURE 4.7 The distribution of National Wildlife Areas and Migratory Bird Sanctuaries across Canada as established by the Migratory Bird Convention Act 1917 and Canadian Wildlife Act 1985. *Source: Environment Canada.*

Some of these laws have provisions for wildlife, such as setting allowable limits of contamination based on wildlife, fish and invertebrate tolerances. Other laws in this category are primarily concerned with human health, but in protecting humans we also protect wildlife. For example, when the US Environmental Protection Agency banned the use of DDT in the United States in 1972, the action said nothing about the effects of DDT on eggshell thinning — only human health was officially considered.[46] However, preventing further application of the pesticide allowed many populations of fish-eating birds to recover.

The *US Clean Air Act of 1970* gave the US Environmental Protection Agency (EPA) the responsibility of establishing standards for air pollution, the authority to enforce these standards, and the ability to place fines and other penalties on violators. The Clean Air Act does not have any wildlife provisions *per se*, except to include them with "soils, water, crops, vegetation, and a short litany of other concerns, but the Act does protect a valuable natural resource — the air we breathe" (Bean 1983[36]; Federal Register 1981[47]). The Clean Air Act was significantly amended several times, with the last major overhaul occurring in 1990.[48] It now has provisions for automobile, aircraft and industrial emissions; fuel efficiency; ozone layer protection; acid deposition or rain; and other air quality factors. Several states have enacted more stringent regulations than those found in the federal law, but they cannot impose less strict rules.

In Canada there have been two Acts entitled "Clean Air". In 1970, the Clean Air Act regulated the atmospheric release of asbestos, lead, mercury and vinyl chloride; it was replaced by the Canadian Environmental Protection Act. The *Canadian Clean Air Act* of 2006 regulated smog and greenhouse emissions. Neither Act was as comprehensive as the Clean Air Act and its amendments in the United States. Like some other Canadian laws on natural resources, clean air and water are shared responsibilities between the federal government and the provinces.

The *Clean Water Act* 1972 in the United States has its roots within the Federal Water Pollution Act 1948. This Act sought "to provide a comprehensive program for preventing, abating, and controlling water pollution".[49] The Act also decided that the primary responsibility for clean water lay with the states, and that federal involvement focused on advice and assistance to the states. No standards were set, and pollution was not specifically prevented.

Under the 1972 Act, the EPA has established wastewater standards, and water quality standards for all contaminants in surface waters; given the lead for clean water, especially in navigable waterways, to the federal government; and set controls on point source pollution. The National Pollutant Discharge Elimination System (NPDES) permit program controls discharges. A focus of the Clean Water Act is navigable waters. The interpretation of this connection has varied through the years. At one time almost all waters, including intermittent streams, lakes, prairie potholes, sloughs and wetlands, were included, and these water bodies received considerable protection. In the 2006 case of *Rapanos v United States*,[50] a majority of the Supreme Court held that the term "waters of the United States ... includes only those relatively permanent, standing or continuously flowing bodies of water forming geographic features" that are described in ordinary parlance as "streams ... oceans, rivers, [and] lakes". Wetlands and shallow bodies of waters totally contained within one state were eliminated from the protective umbrella of the Act. The Act has become extremely complex over the years but it does provide for fish and wildlife through regulations on the minimum dissolved oxygen concentrations in natural waters and by recognizing wildlife in oil spills or dumping of dredged materials.[51]

In Canada, waters that fall entirely within the confines of a single province are under the jurisdiction of that province.[52] Thus, each of the provinces has its own Clean Water Act. We don't have the space to cover each of these laws, so if you're interested, do what I would do: Google it.

In 2000, the Canadian Parliament passed an umbrella environmental protection bill called the *Canadian Environmental Protection Act* (CEPA) 1999. The guiding principles of the Act, according to Environment Canada,[53] are as follows.

- *Encourage Sustainable Development* – the Government of Canada's environmental protection strategies are driven by a vision of environmentally sustainable economic development.
- *Pollution Prevention* – CEPA 1999 shifts the focus away from managing pollution after it has been created, to preventing pollution.
- *Virtual Elimination* – CEPA 1999 requires the virtual elimination of releases of substances that are persistent (take a long time to break down), bioaccumulative (collect in living organisms and end up in the food chain), toxic (according to CEPA 1999 Section 64) and primarily the result of human activities.
- *Ecosystem Approach* – based on natural geographic units rather than political boundaries, the ecosystem approach recognizes the interrelationships between land, air, water, wildlife and human activities.
- *Precautionary Principle* – the government's actions to protect the environment and health are guided by the precautionary principle, which states that "where there are threats of serious or irreversible damage, lack of full scientific certainty shall not be used as a reason for postponing cost-effective measures to prevent environmental degradation."
- *Intergovernmental Cooperation* – CEPA 1999 reflects that all governments have the authority to protect the environment, and directs the federal government to endeavor to act in cooperation with governments in Canada to ensure that federal actions are complementary to and avoid duplication with other governments.

- *National Standards* – CEPA 1999 reinforces the role of national leadership to achieve ecosystem health and sustainable development by providing for the creation of science-based, national environmental standards.
- *Polluter Pays Principle* – CEPA 1999 embodies the principle that users and producers of pollutants and wastes should bear the responsibility for their actions.
- *Science-Based Decision-Making* – CEPA 1999 emphasizes the integral role of science and traditional aboriginal knowledge (where available) in decision-making and that social, economic and technical issues are to be considered in the risk management processes.

The law sets standards for contaminant concentrations in the air, water and land for the primary protection of humans and non-human organisms.

There is one last law that I would like to cover under this section of environmental protection. It's different from the Clean Air and Water Acts, for it is not really regulatory but procedural – which means that federal agencies have to comply with the procedural rules of the Act rather than any standards. This is the *National Environmental Policy Act* (NEPA) of 1970.[54] This Act has been called the "Magna Carta of Environmental Legislation"[55] because, like the Magna Carta of England which helped level the playing field between the monarchy and the nobility, NEPA levels the field between the environment and other federal priorities. NEPA requires that every major federal activity involving the land or environment undergoes an environmental review. These reviews may consist of an *Environmental Assessment* (EA) for minor projects or *Environmental Impact Statements* (EIS) for larger projects. An EA can be a few to several pages long, indicating that various factors were considered but logic and science would dictate that they will not impact the environment. An EIS, however, is much more extensive, usually involving one or more on-site surveys and detailed analyses on how the project might impact the environment.[56] The other factor required by NEPA is for the public to be notified and have an opportunity to respond to the planned activity. Thus, federal activities have become much more transparent than they were some 50 years ago.

Of course, far-reaching Acts such as this one always has their detractors. Criticisms of NEPA include the arguments[55,57–59] that:

- Its enforcement has focused too much on "jumping through the hoops" [my phrase] and not on decisions based on high quality science and information
- The public have not been informed early enough in the process
- Alternatives have not been sufficiently explored
- Common decision processes such as risk analysis are not adequately applied
- The courts have not always made decisions objectively, and on occasion the courts seem to be enacting their own legislation.

However, these complaints involve only a fraction of all NEPA cases. Compared to allowing agencies to do whatever they please, and preventing the American public from knowing about projects until it is too late, the Act brings sunshine into what used to be dark and mysterious processes.

In Canada, the *Environment Assessment Agency* performs many of the functions conducted by NEPA. Except for actions involving the Canadian Nuclear Safety Commission or the National Energy Board, it determines when environmental assessments need to be made, assists in the process, evaluates assessments when they are completed, and assures that the mitigation efforts identified in the assessments are being carried out and are effective. The agency receives its authority through the Canadian Environmental Assessment Act 2012.

Study Questions

4.1. What is the most contentious policy issue in the US Endangered Species Act of 1973? Why do you suppose its enforcement has varied through the years?

4.2. Similarly, the Canadian Species at Risk Act received considerable resistance by parliament during its long struggle to be enacted. What were some of the driving forces behind this resistance? Do these factors still exist in Canadian politics?

4.3. What are the benefits of greater transparency in agency activities that were provided by Acts such as the National Environmental Policy Act in the United States and the Canadian Environmental Assessment Act?

4.4. As we have seen in Chapter 2 and in this chapter, both Canada and the United States have passed similar laws to protect the environment. Discuss with your class and instructor why it appears at least that Canada has been about 15−20 years later than the United States in enacting these laws.

REFERENCES

1. Toner G, Meadowcraft J. The struggle of the Canadian federal government to institutionalize sustainable development. In: VanNijnatted DL, Boardman R, eds. *Canadian Environmental Policy and Politics*. 3rd Ed. New York, NY: Oxford University Press; 2009:77−90.
2. Bean MJ, Rowland MJ. *The Evolution of National Wildlife Law*. Westport, CT: Praeger; 1997.
3. Schwartz M. The performance of the endangered species act. *Ann Rev Ecol Evol Syst*. 2008;39:279−299.
4. Bean MJ. *The Evolution of National Wildlife Law*. New York: Environment Defense Fund, Praeger; 1983.
5. US Fish and Wildlife Service. 2011. A history of the endangered species act of 1973. US Fish and Wildlife Service, Arlington VA. <http://www.fws.gov/endangered/esa-library/pdf/history_ESA.pdf>.
6. Federal Register (*ESA*; 7 U.S.C. § 136, 16 U.S.C. § 1531 *et seq.*).
7. Taylor MFJ, Suckling KF, Rachlinski JJ. The effectiveness of the endangered species act: a quantitative analysis. *BioScience*. 2005;55:360−367.

8. Hagen AM, Hodges KE. Resolving critical habitat designation failures: reconciling law, policy, and biology. *Conserv Biol.* 2006;20:399–407.

9. US Fish and Wildlife Service. <http://ecos.fws.gov/speciesProfile/profile/speciesProfile. action?spcode=D02D>.

10. *Federal Register*/Vol. 75, No. 51/Wednesday, March 17, 2010/Rules.

11. Fact Sheet. The Northern Spotted Owl. Defenders of Wildlife.

12. US Fish and Wildlife Service. Northern Spotted Owl information site. <http://www.fws. gov/oregonfwo/Species/Data/NorthernSpottedOwl/main.asp>; 2012.

13. Wiens JD, Anthony RG, Forsman ED. Barred owl occupancy surveys within the range of the northern spotted owl. *J Wildl Manage.* 2011;75:531–583.

14. Mann CC, Plummer ML. *Noah's Choice: The Future of Endangered Species.* New York, NY: Alfred Knopf; 1995.

15. Pombo RW. *The ESA at 30: Time for Congress to update and strengthen the law.* US House of Representatives Resources Committee; 2004.

16. Schwartz MW. Choosing the appropriate scale of reserves for conservation. *Ann Rev Ecol System.* 1999;30:83–1008.

17. Male TD, Bean MJ. Measuring progress in US endangered species conservation. *Ecol Lett.* 2005;8:986–992.

18. US Fish and Wildlife Service. Federal and state threatened and endangered species expenditures, fiscal year 2004. Washington, DC; 2006.

19. US Fish and Wildlife Service. Federal and state endangered and threatened species expenditures. Washington, DC. <http://www.fws.gov/endangered/esa-library/pdf/2011.EXP.final. pdf>; 2011.

20. Shogren JF, Tschirhart J, Anderson T, et al. Why economics matters for endangered species protection. *Conserv Biol.* 1999;13:1257–1261.

21. Roughgarden J. Can economics save biodiversity? In: Swanson T, ed. *The Economics and Ecology of Biodiversity Decline: The forces driving global change.* New York, NY: Cambridge University Press; 1995:149–153.

22. Ralston H. III. Feeding people versus feeding saving nature? In: Light A, Rolston III H, eds. *Environmental Ethics: an anthology.* Malden, MA: Blackwell Publishing; 2003:451–462.

23. Martin F, Taylor J, Suckling KF, Raschlinski JJ. The effectiveness of the endangered species act: a quantitative analysis. *BioScience.* 2005;55:360–367.

24. National Oceanic and Atmospheric Administration. Marine Mammal Protection Act. <http://www.nmfs.noaa.gov/pr/laws/mmpa/>; 2013.

25. National Oceanic and Atmospheric Administration. National Marine Fisheries Service. <www.nmfs.noaa.gov>; 2013.

26. Elgie S. The politics of extinction: the birth of Canada's species at risk act. In: VanNijnatten DL, Boardman R, eds. *Canadian Environmental Policy and Politics Prospects for Leadership and Innovation.* Don Mills, ON: Oxford University Press; 2009:197–215.

27. Committee on the Status of Endangered Wildlife in Canada. About COSEWIC. <http:// www.cosewic.gc.ca/eng/sct6/sct6_3_e.cfm#hist>; 2013.

28. Environment Canada. <http://www.ec.gc.ca>; 2013.

29. Environment Canada. Habitat Stewardship Program for Species at Risk. <http://www.ec. gc.ca/hsp-pih/default.asp?lang=En>; 2013.

30. Environment Canada. The Wild Animal and Plant Protection and Regulation of International and Interprovincial Trade Act (WAPPRIITA). <http://www.ec.gc.ca/alef-ewe/default.asp?lang=en&n=65FDC5E7-1>; 2013.

31. Wild Animal and Plant Protection and Regulation of International and Interprovincial Trade Act S.C. 1992; c. 52.

32. Canada, Parliament, Report of the standing committee on environment. A Global Partnership. 1993; April: 30.

33. Committee on International Trade in Endangered Species. <www.cites.org>.

34. International Union for the Conservation of Nature. <www.icun.org>.

35. US Fish and Wildlife Service. International Affairs. The Lacey Act. <http://www.fws.gov/international/laws-treaties-agreements/us-conservation-laws/lacey-act.html>; 2013.

36. Bean MJ. *The Evolution of National Wildlife Law. Revised and Expanded Edition.* New York, NY: Praeger; 1983.

37. Federal Register (16 U.S.C. 703).

38. Federal Register, (16 U.S.C. 703−712; Ch. 128; July 13, 40 Stat. 755) plus amendment <http://www.fws.gov/laws/lawsdigest/migtrea.html>; 1918.

39. Wiemeyer SN, Mulhern BM, Ligas FJ, et al. Residues of organochlorine pesticides, polychlorinated biphenyls, and mercury in bald eagle eggs and changes in shell thickness − 1969 and 1970. *Pestic Monit J.* 1972;6:50−55.

40. Wiemeyer SN, Lamont TJ, Bunck CM, Sindelar CR, Gramlich FJ, Fraser JD, Byrd MA. Organochlorine pesticide, polychlorobiphenyl, and mercury residues in bald eagle eggs 1969−79 and their relationships to shell thinning and reproduction. *Arch. Environ. Contam. Toxicol.* 1984;13:529−549.

41. Grubb TG, Wiemeyer SN, Kiff LF. Eggshell thinning and contaminant levels in bald eagle eggs from Arizona, 1977−1985. *Southwest Natur.* 1990;35:298−301.

42. Federal Register (16 U.S.C. 668(a); 50 CFR 22), (16 U.S.C. 668c; 50 CFR 22.3).

43. The Wild Free-Roaming Horses and Burros Act of 1971. (Public Law 92−195.)

44. Bureau of Land Management. National Wild Horse and Burro Program. <http://www.blm.gov/wo/st/en/prog/whbprogram.html>; 2013.

45. Environment Canada. 2013. Migratory Bird Conventions Act. 1994. <http://www.ec.gc.ca/alef-ewe/default.asp?lang=en&n=3DF2F089-1>; 2013.

46. US Environmental Protection Agency. DDT Ban Takes Effect <http://www2.epa.gov/aboutepa/ddt-ban-takes-effect>; 2013.

47. Federal Register. *Part 50 − National Primary and Secondary Ambient Air Quality Standards.* Washington, DC: Office of the Federal Register; 1981.

48. US Environmental Protection Agency. Air pollution and the Clean Air Act. <http://www.epa.gov/air/caa/>; 2013.

49. Federal Water Pollution Control Act (FWPCA) (P.L. 80−845, 62 Stat. 1155).

50. Federal Register 547 U.S. 715. 2006.

51. US Fish and Wildlife Service. Federal Register. Digest of Federal Resource Laws of Interest to the US Fish and Wildlife Service Federal Water Pollution Control Act (Clean Water Act). <http://www.fws.gov/laws/lawsdigest/FWATRPO.HTML>; 2013.

52. Environment Canada. <https://www.ec.gc.ca/eau-water/default.asp?lang=En&n=24C5BD18-1>; 2013.

53. Environment Canada. A Guide to Understanding the Canadian Environmental Protection Act, 1999. <https://www.ec.gc.ca/lcpe-cepa/E00B5BD8-13BC-4FBF-9B74-1013AD5FFC05/Guide04_e.pdf>; 2004.

54. 42 U.S.C. Section 4321.

55. Broussard SR, Whitaker BD. The Magna Carta of environmental legislation: A historical look at 30 years of NEPA − forest service litigation. *Forest Policy Econ.* 2009;11:134−140.

56. US Environmental Protection Agency. National Environmental Protection Act. Basic Information. <http://www.epa.gov/compliance/basics/nepa.html>; 2012.

57. Malmsheimer R, Floyd D. US Courts of appeals judges' review of federal natural resource agencies' decisions. *Soc Natur Resour*. 2004;17:533–546.

58. Salk MS, Tolbert VR, Dickerman JO. Guidelines and techniques for improving the NEPA process. *Environ Manage*. 1997;23:467–476.

59. Fairbrother A, Turnley JC. Predicting risks of uncharacteristic wildfires: application of the risk assessment process. *Forest Ecol Manage*. 2005;211:28–35.

Part III

The Bureaucracy of Natural Resources

Chapter 5

Federal Administration in Canada

Terms to Know

- Constitutional Monarchy
- Constitution Act 1982
- Environment Canada
- Environmental Stewardship Branch
- Canadian Wildlife Service
- Meteorological Branch
- Science and Technology Branch
- Dominion Wildlife Service
- Department of the Environment Act 1971
- International Rivers Improvement Act 1955
- Federal Sustainable Development Act 2008
- Environmental Enforcement Act 2010
- Parks Canada
- Dominion Forest Reserves and Parks Act 1911
- Geological Survey
- NRCan
- Earth Sciences Sector
- Energy Sector $$$
- Minerals and Metals Sector
- Canadian Forest Service
- Arctic Waters Pollution Prevention Act 1985
- Canada Oil and Gas Operations Act 1985
- Canada Petroleum Resources Act 1985
- Department of Natural Resources Act 1995
- Forestry Act 1989
- Nuclear Energy Act 1985
- Ecosystems and Fisheries Sector
- Ecosystems and Oceans Science
- Canada Shipping Act 2001
- Coastal Fisheries Protection Act 1985
- Department of Fisheries and Oceans Act 1985
- Fisheries Act 1982.
- Oceans Act 1996 (amended 2005)

D.W. Sparling: Natural Resources Administration. DOI: http://dx.doi.org/10.1016/B978-0-12-404647-4.00005-2
105

INTRODUCTION

The Canadian government is a *Constitutional Monarchy*.[1] On paper, all legislation at the federal and provincial levels needs to be approved by the monarchy, i.e. the King or Queen of England as represented by the Governor General for the federal government or lieutenant governors for provinces. Additionally, the monarchy has the authority to appoint a prime minister over the national government, name premiers over provinces, and dissolve parliament if deemed necessary. In practice, however, the monarchy is largely symbolic with automatic approval of all legislation and popular elections replacing monarchical decisions in determining leadership. The chief elected official of the federal government is the prime minister, who arguably has broader internal power than President of the United States. Unless he or she loses the confidence of parliament and the people, the prime minister can serve for life. The prime minister controls the appointments of many key figures in Canada's government, including the Cabinet members, Supreme Court justices, senators, heads of crown corporations, ambassadors to foreign countries, provincial lieutenant governors, and over 3000 other offices. Further, the prime minister plays a prominent role in the legislative process, with the majority of bills put before parliament originating in the Cabinet, and is the commander of the Canadian Forces.

Other than that, the Canadian federal government has many of the same overall responsibilities as those in the United States plus a few more according to their *Constitution Act* 1982. The Canadian government has primary jurisdiction over any form of taxation; international/interprovincial trade and commerce, communications and transportation; banking and currency; foreign affairs (treaties); militia and defense (there is no equivalent to a state national guard); criminal law and penitentiaries (provinces have short term jails for terms of two years or less, but no long-term prisons); and naturalization; and works with the First Nations (Indian reservations in the United States).[2-4]

There is at least one major difference between the two nations in the way that natural resources are handled. Since at least 1900, state and federal laws have coexisted to manage natural resources in the United States. Often the federal government has taken the lead — as, for example, with the Clean Air Act, Clean Water Act, Endangered Species Act, and so forth. States then often adopt their own laws in concert with the federal regulations. In many cases states can impose stricter laws than those provided by the federal government, but they cannot relax federal regulations. Fisheries and wildlife management are somewhat different, in that the states provide the basis for the Public Trust Doctrine and the federal government has less to say about wildlife and fisheries — the obvious exception being endangered species.

In Canada, division of powers between provinces and the nation (or Parliament) accords the provinces with primary responsibility over many of the natural resource issues; when it comes to issues affecting the entire nation, however, the federal agencies may have jurisdiction. As a result, the federal government has sometimes been criticized as being sluggish in becoming involved in the management of natural resources, even when the issues have an international basis.[5–7] This can be seen, for example, in the apparent reluctance of the federal government to develop the Species at Risk Act to protect endangered species (see Chapter 4). Note that the difference between the two nations is more of a matter of degree in the division of authority between federal and lower government units than an actual qualitative difference. In the United States, the federal government takes a stronger role in environmental issues than in Canada. However, in both nations there is shared responsibility between federal and state or provincial jurisdictions.

The cabinet of the Executive Branch of the Canadian government, also called the Canadian Ministry, includes approximately 40 separate ministries and 5 offices.[8] The heads of these ministries are also members of parliament and can introduce legislation. Among these ministries, three have primary influence on natural resources and one, the Ministry of Agriculture, has programs on greenhouse gases and pesticides. The three principal ministries, Environment Canada, Natural Resources Canada, and Fisheries and Oceans, will occupy the rest of this chapter.

ENVIRONMENT CANADA

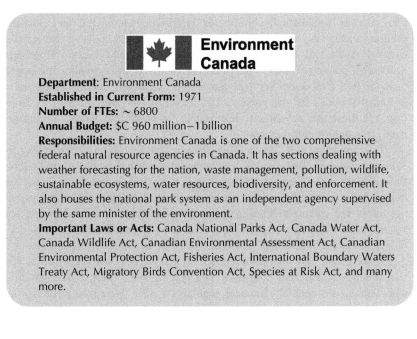

Department: Environment Canada
Established in Current Form: 1971
Number of FTEs: ~ 6800
Annual Budget: $C 960 million–1 billion
Responsibilities: Environment Canada is one of the two comprehensive federal natural resource agencies in Canada. It has sections dealing with weather forecasting for the nation, waste management, pollution, wildlife, sustainable ecosystems, water resources, biodiversity, and enforcement. It also houses the national park system as an independent agency supervised by the same minister of the environment.
Important Laws or Acts: Canada National Parks Act, Canada Water Act, Canada Wildlife Act, Canadian Environmental Assessment Act, Canadian Environmental Protection Act, Fisheries Act, International Boundary Waters Treaty Act, Migratory Birds Convention Act, Species at Risk Act, and many more.

Environment Canada (EC) was pieced together in 1971 from other elements of the Canadian federal government, including the Meteorological Service, Canadian Wildlife Service, Fisheries Service and others.

As quoted from the agency website[9]:

Environment Canada's mandate is to:

- *Preserve and enhance the quality of the natural environment, including water, air, soil, flora and fauna;*
- *Conserve Canada's renewable resources;*
- *Conserve and protect Canada's water resources;*
- *Forecast daily weather conditions and warnings, and provide detailed meteorological information to all of Canada;*
- *Enforce rules relating to boundary waters; and*
- *Coordinate environmental policies and programs for the federal government.*

Organizational Structure of Environment Canada

In addition to provincial or regional offices, EC has several branches that carry out these services (Figure 5.1). Somewhere in their organizational structure, all federal and provincial/state natural resource agencies have three types of offices: (1) those that deal directly with the resource(s); (2) those that run the everyday operations of the agency; and (3) those that provide specific technical assistance. For example, the Strategic Policy Office in EC develops the direction the agency will go, assesses if it is adhering to guidelines already established, coordinates interagency activities, and provides economic analyses. The General Council is the legal branch of the agency, General Audit and Evaluations oversees finances, and Human Resources deals with personnel issues; all three are involved with running the operations of EC. Corporate Services and Communications contain elements of technical services for the agency. The branches that most interest us, however, are those that are directly responsible for natural resources, in management, research or enforcement.

The *Environmental Stewardship Branch* oversees both enforcement for wildlife resources and overall environmental enforcement for the nation. Enforcement covers the manufacture and use of toxic substances, and the import and export of hazardous wastes and materials; laws concerning migratory birds and endangered species; and the protection of domestic water and water shared internationally. EC also contains the *Canadian Wildlife Service* (CWS), which works with national interests in endangered species and migratory birds, and coordinates international agreements on wildlife. The stewardship branch also houses offices dealing with energy, transportation and with environmental chemical regulations. The *Meteorological Service* handles dissemination of climate and weather information for the country. *Science and Technology* is the research and development arm of EC, and

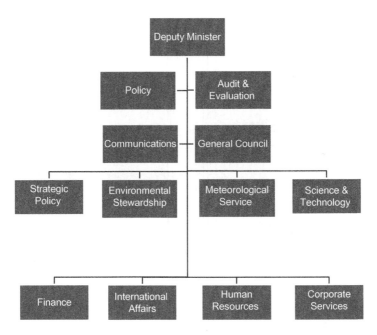

FIGURE 5.1   Organizational structure of Environment Canada. As with all departments of ministries, there are offices for technical services, running the operations of the agency, and establishing policy, and some that actually deal with the natural resources for which they have charge. *Data: Environment Canada.*

conducts research on global climate change, wildlife and landscape ecology, environmental risk assessment, and water resources. Environment Canada operates 15 research institutes, 7 storm prediction centers, and 32 water survey offices scattered throughout the country. It also maintains extensive air, climate, and water monitoring networks.

More on the Canadian Wildlife Service

Although the CWS is a department totally within EC, we would be remiss if we did not spend some space discussing this very important agency. The CWS predates Environment Canada by about 24 years, and was created in 1947 as the primary federal wildlife agency in Canada. It is analogous to the US Fish and Wildlife Service and is responsible for the federal species at risk (SARA) program, protection and management of migratory birds, and management of the National Wildlife Areas and Migratory Bird Sanctuaries that provide protection for wildlife across the nation (Figure 5.2). It also deals with international agreements on species, such as the Migratory Bird Treaty Act with the United States and The Convention on International Trade in Endangered Species of Wild Flora and Fauna (CITES).

FIGURE 5.2 Cap Tourmente National Wildlife Area, Quebec, Canada. This is one of the 54 wildlife areas managed by the Canadian Wildlife Service. *Credit: Boreal through Wikipedia Commons.*

The Canadian Wildlife Service has its roots in the early 1900s, when naturalists in both the United States and Canada were growing concerned about the diminishing wildlife and other natural resources in both countries. England (representing Canada) and the United States signed the joint Migratory Bird Convention Act in 1917. In 1947 the precursor to the CWS, the *Dominion Wildlife Service*, was established by Parliament, and three years later this agency's name was changed to the current Canadian Wildlife Service. In 1973 the CWS was directed to enforce the Canada Wildlife Act, which gave the federal government authority to conduct wildlife research and to work with the provinces in wildlife conservation and interpretation. In 2003, funding and direction for CWS shifted towards endangered and threatened species and away from migratory birds. The department was also substantially downsized in 2010 as many of its functions in research, information transfer and enforcement were stripped away. Currently, the CWS employs approximately 450 staff.[10]

Funding of Environment Canada Programs

According to EC's website, over two-thirds of its budget and more than half of its workforce are dedicated to science and technology. Canadian federal budgets are focused more on thematic issues than on specific agencies, thus it is difficult to determine, for example, how much of the budget the Canadian Wildlife Service actually gets. We can see, however, that climate change research, general operations and weather or meteorology receive the greatest share of funding for the agency (Figure 5.3).[11] The budget for the CWS comes from the biodiversity and ecosystems accounts.

Percent

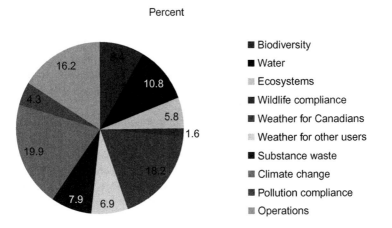

- ■ Biodiversity
- ■ Water
- ▨ Ecosystems
- ■ Wildlife compliance
- ■ Weather for Canadians
- ▨ Weather for other users
- ■ Substance waste
- ■ Climate change
- ■ Pollution compliance
- ▨ Operations

FIGURE 5.3   The annual budget in terms of percent of total budget for Environment Canada. *Data: Treasury Board of Canada Secretariat. This figure is reproduced in color in the color plate section.*

Acts Administered by Environment Canada

The Canadian Parliament has charged Environment Canada with administering several Acts associated with environmental protection and wildlife. Some of the more comprehensive of these Acts include the following.[12]

> *The Department of the Environment Act* 1971. This Act established EC as a department under the Minister of the Environment, and gave it the responsibility to preserve and enhance the quality of the natural environment, provide meteorological services, and coordinate policies and programs to achieve environmental objectives.
> *Clean Water Act* 1985 and *International Rivers Improvement Act* 1955. The Clean Water Act provides for the management of water resources, including research and planning and implementation of conservation measures. The International Rivers Improvement Act is an older but still viable law that seeks to maintain and improve watersheds of rivers that flow between the United States and Canada, such as the St Lawrence and many that exist in the western provinces.
> *Canadian Environmental Protection Act* 1999. This Act is the cornerstone of Canada's environmental legislation and an important part of Canada's broader legislative framework aimed at preventing pollution and protecting the environment and human health. It is discussed in greater detail in Chapter 4.
> *Species at Risk Act (SARA)* 2002. This Act, also discussed in Chapter 4, is the Canadian counterpart to the US Endangered Species Act. It is a powerful law that protects species at risk, maintains healthy ecosystems and preserves Canada's natural heritage.

Migratory Birds Convention Act 1917. This law is Canada's counterpart to the US Migratory Bird Treaty Act signed with the United States in 1918. It was extensively revised in 1994. It provides protection for all of Canada's migratory birds, and more details can be gleaned in Chapter 4. Notably, the Migratory Birds Convention Act allowed the formation of migratory bird sanctuaries throughout Canada with a total area of around 28.4 million acres (11.5 million ha).

Wild Animal and Plant Protection and Regulation of International and Interprovincial Trade Act 1992. This law was passed as a formal commitment to Canada's involvement with the Convention on International Trade in Endangered Species of Wild Flora and Fauna (CITES) to control illegal trade in species of concern and to protect Canadian ecosystems from the introduction of harmful exotic species. See Chapter 4 for further information.

Canada Wildlife Act 1973. The Canada Wildlife Act allows the creation of National Wildlife Areas. It also provides guidance on what activities can be conducted on these areas. Currently there are 146 sites encompassing around 2.5 million acres (1 million ha), nearly half of which are in marine areas.[13]

Federal Sustainable Development Act 2008. The Minister of the Environment is responsible for developing the Federal Sustainable Development Strategy, setting out goals and targets for all federal departments to improve the standard of living by protecting human health, conserving the environment, using resources efficiently, and advancing long-term economic competitiveness. See Chapter 1 for more about sustainability. The first federal strategy was adopted in 2010.

Canadian Environmental Assessment Act 2012. This Act and the agency it formed, the *Canadian Environmental Assessment Agency*, are discussed in Chapter 4. The agency oversees all of the environmental assessments for the federal government, and is an independent agency under the auspices of EC.

Environmental Enforcement Act 2010. This Act amended several of the above Acts to provide a common set of principles, factors and penalties to be taken into account in sentencing in actions contrary to these Acts. It provides a more formalized process for protecting Canada's National arks, air, land, water, and wildlife. It was initially passed in 2010, and has been gradually phased in over three years.

Parks Canada

Parks Canada is the agency responsible for the national parks, historic sites and marine conservation areas within Canada. Although it resides under the EC umbrella, it, like the Environmental Assessment Agency, is considered to be an independent agency supervised by the Minister of the Environment but not within the organizational line with the rest of EC. Parks Canada manages 37 national parks, 7 national park reserves, 4 national marine conservation areas, the Pingo National Landmark on the shores of the Beaufort Sea, 167 national historic sites, and the Canadian Register of Historic Places.

Like those in the United States, Canada's national parks and areas provide some of the most breathtaking and awesome landscapes in the world. Few sights are as memorable as seeing a golden eagle (*Aquila chrysaetos*) soaring above the rugged landscape of Banff National Park in Alberta (Figure 5.4), or the 17-m (56 ft) tides in Fundy National Park in New Brunswick. The national park system began in a similar way as in the United States. As conservationists began to see the loss of natural areas in the late 1800s, they undertook measures to preserve some of the most amazing of these sites. The construction of the first transcontinental railways helped in this process, for they offered a way for people to observe firsthand some of the natural treasures both nations had to offer. The first national park in Canada was Banff, also called Rock Mountain Park during part of its history. Hot springs in the region were initially placed under protection in 1885 and, through the efforts of Thomas White, Canada's Minister of the Interior, *the Dominion Forest Reserves and Parks Act* was given Royal assent in 1911; along with it Banff National Park was born. Both economic incentives and preservation were motives for the first national parks as the nation encouraged development of hotels and other hospitality businesses within them. In the earliest days commercial

FIGURE 5.4   Few places on earth are as beautiful as the rugged landscape of Canada's first national park, Banff National Park in Alberta.

development, including resource extraction in timber, mining, and other purposes, was conducted in Canada's national parks. This was in direct contrast to those in the United States, whose main purposes were preservation and tourism. It was not until 1930, with the passing of the Parks Act, that mining, mineral exploration and development were banned. Limited forestry practices were allowed if they were compatible with maintaining the environmental health of the parks. While the need for restoration of the parks was widely acknowledged, funding and other incentives were slow in coming until the Canada National Parks Act of 2001 reinforced the necessity of maintenance and restoration of ecological integrity by saving natural resources and ecosystem.[14]

Today, the national parks stretch from one coast to the other and from the border with the United States into the Arctic (Figure 5.5). The mandate for the agency[15] is:

we protect and present nationally significant examples of Canada's natural and cultural heritage and foster public understanding, appreciation and enjoyment in ways that ensure their ecological and commemorative integrity for present and future generations.

Nowhere in their charter is there any statement about using or exploiting resources. The Canadian National Parks today are for the pleasure and education of visitors, not for making a profit. This is easily seen in that more than 70% of the approximately more than C$597 million is applied to visitor experiences, heritage resource conservation, and public appreciation (Figure 5.6).[16]

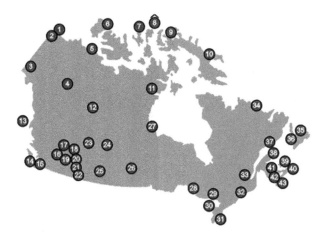

FIGURE 5.5 The National Park system of Canada stretches from one coast to the other and from the border with the United States to the Arctic. *Credit: Trailcanada.com.*

Percent

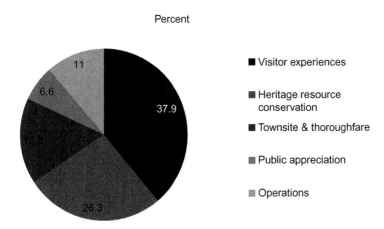

- ■ Visitor experiences
- ■ Heritage resource conservation
- ■ Townsite & thoroughfare
- ■ Public appreciation
- ▨ Operations

FIGURE 5.6 The annual budget in terms of percent total budget for Natural Resources Canada. *Data: Treasury Board of Canada Secretariat. This figure is reproduced in color in the color plate section.*

NATURAL RESOURCES CANADA

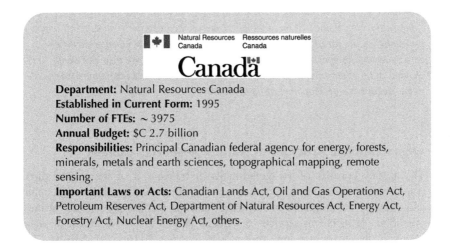

Natural Resources Canada   Ressources naturelles Canada

Canada

Department: Natural Resources Canada
Established in Current Form: 1995
Number of FTEs: ~ 3975
Annual Budget: $C 2.7 billion
Responsibilities: Principal Canadian federal agency for energy, forests, minerals, metals and earth sciences, topographical mapping, remote sensing.
Important Laws or Acts: Canadian Lands Act, Oil and Gas Operations Act, Petroleum Reserves Act, Department of Natural Resources Act, Energy Act, Forestry Act, Nuclear Energy Act, others.

Natural Resources Canada (NRCan) has early roots. In 1842, 25 years before Canada was united under Confederation, the ancestor of NRCan, *the Geological Survey*, was enacted. The Survey was established to locate and catalog commercially useful sources of minerals; it was also commissioned to catalog water sources and soil types of Canada, which at that time consisted only of Quebec and southern Ontario.[17]

Like the early geologists in the United States (see Chapter 6), the Canadian scientists did a great service in exploring unknown areas of Canada, and their efforts increased as provinces were added to the nation. The field geologists brought back information on the flora and fauna, and Native Americans, from unexplored areas. Forestry research began in the 1880s, as scientists for the survey took specimens of many trees during these early years and studied the lumber quality of the various woods. Mapping was added to the duties of the Geological Survey in 1854 through the urging of its founder, William Logan. In 1907, Parliament officially added the responsibility of mapping forested areas of Canada to the agency's list of duties. These efforts were formalized into the Topographical Division of the Survey in 1908. Natural Resources Canada became official in 1995, when Parliament combined the departments of energy, mines and minerals, and forests. At that time, the old Geological Survey was also incorporated into NRCan.

The mandate or mission statement of NRCan, as gleaned from its website,[18] is:

to enhance the responsible development and use of Canada's natural resources and the competitiveness of Canada's natural resources products. We are an established leader in science and technology in the fields of energy, forests, and minerals and metals and use our expertise in earth sciences to build and maintain an up-to-date knowledge base of our landmass. NRCan develops policies and programs that enhance the contribution of the natural resources sector to the economy and improve the quality of life for all Canadians. We conduct innovative science in facilities across Canada to generate ideas and transfer technologies. We also represent Canada at the international level to meet the country's global commitments related to the sustainable development of natural resources.

Organizational Structure of Natural Resources Canada

The agency consists of 16 departments or bureaus; 4 of these have direct responsibility for natural resources, several others are involved with policy and strategic development of resources, and the rest are part of the business operations or technical services of NRCan (Figure 5.7). The *Earth Sciences* sector is the mapping and information transfer division of NRCan. The *Energy Sector* advises the government on federal energy policies, strategies, emergency plans and activities, and helps to assure energy conservation. *Minerals and Metals* is really the old Geological Survey with some added functions. This sector is the primary source of technological, scientific and policy expertise on mineral and metal resources, and on explosives regulation and technology. It contains three major research laboratories that focus on minerals, explosives, and materials technology.

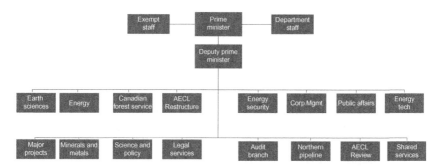

FIGURE 5.7 Organizational structure of Natural Resources Canada. *Data from Natural Resources Canada.*

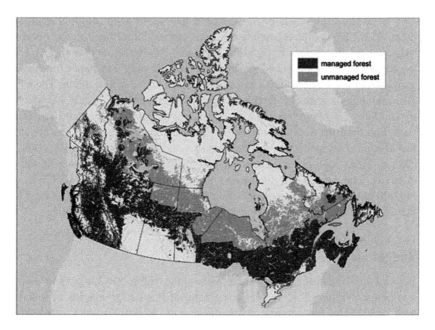

FIGURE 5.8 Distribution of forests throughout Canada. Due largely to the boreal forests, Canada is among the top 4 countries in the world in terms of amount of forest, much of which is unbroken. *Credit: Natural Resources Canada. This figure is reproduced in color in the color plate section.*

The *Canadian Forest Service* (CFS), which is also a part of NRCan, has very important natural resource functions. Due to its extensive boreal forests, Canada is among the top 10 countries in the world in forested area.[19] The country contains 766 million acres (310 million ha) of forested land, 94% of which is publically owned and substantially unfragmented (Figure 5.8). An estimated two-thirds of Canada's 140,000 species of plants and animals, including 180 different species of trees reside, in these forests. Moreover, 20% of the world's freshwater is found in these forests, and the streams of the East and West coasts provide habitat for much of the continent's salmon reproduction.

However, not all of these forests are under federal jurisdiction. In fact, there is no National Forest System as there is in the United States. All forests within provinces are managed by the respective province. Even the broad expanses of forest in the northern territories south of the Arctic, although Crown lands, are managed by the territories under guidance from the CFS. The Canadian Forest Service "is a *science-based policy organization* within Natural Resources Canada, a Government of Canada department that helps shape the natural resources sector's important contributions to the economy, society and the environment"[20] (italics my own). Therefore, the duties of the CFS lie more in the areas of research and innovation than in direct management. The CFS conducts studies on forest regeneration, health, and optimal management strategies. It provides guidance in marketing the timber and improving the efficiency and sustainability of the pulp and paper industries, and assists First Nations in managing the forests under their domain. The CFS also maintains a national database with forest information and is a leader in the country in remote sensing and monitoring of forest conditions. The Canadian Forestry Service operates seven research centers across the country that specialize in many types of forest research, including insect pests, climate change, forest ecology, ecogenomics, and forest ecosystem dynamics and productivity. The Acadia and Petawawa research forests, comprising 46,950 acres (19,000 ha) in eastern Canada are the only forests that are solely managed by the CFS, and they serve as "living laboratories" for research purposes.

Budget for NRCan

Beyond any doubt, the largest share of funding within NRCan goes to global competition (Figure 5.9). Eighty-six percent of this sector consists of mandatory obligations for ongoing program support. It also includes offices for marketing innovations for the sale of Canadian natural resources and investments in natural resource sectors. The next largest expense is environmental responsibility, which includes research and development in healthy ecosystem initiatives. Information management and agency operations are comparatively small expenses.

Acts Administered by Natural Resources Canada

NRCan is a busy ministry with 31 federal Acts currently under its authority. In this section, however, we will cover only a select few of the most important Acts.

Percent

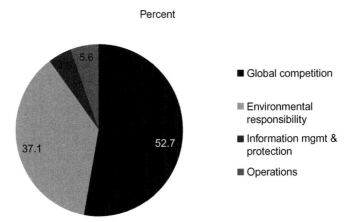

FIGURE 5.9   Annual budget for Natural Resources Canada in terms of percent total budget. *Data: Treasury Board of Canada Secretariat. This figure is reproduced in color in the color plate section.*

Arctic Waters Pollution Prevention Act 1985. The Arctic includes around 40% of Canada's landmass and about 100,000 people. The consistently low temperatures of its waters, ice masses and land create unusual conditions for any contaminants that may be found there, and while many of these pollutants would break down in more temperate climates they often last for years in the Arctic. This Act seeks to protect the waters surrounding the Canadian Arctic from such contamination.

Canada Oil and Gas Operations Act 1985. This Act covers the exploration, production, processing and transportation of oil and gas in federal maritime areas up to 12 nautical miles beyond the coastline. Its purpose is to promote safety, protection of the environment, conservation of oil and gas resources, and joint production agreements.

Canada Petroleum Resources Act 1985. This Act regulates the lease of federally owned oil and gas rights for exploration and development on "frontier lands" that include the "territorial sea" – 12 nautical miles beyond the outer coastline and beyond. It allows the federal government to give permission for oil and gas exploration to occur on frontier lands, and it provides the opportunity for the government to protect the environment by attaching exploration restrictions when leasing rights or by stopping work if there is an environmental problem.

Department of Natural Resources Act 1995. This Act created the Natural Resources Canada ministry and defined its mission. The Act gave the agency specific powers over natural resources, explosives and technical surveys not already assigned to Environment Canada or Fisheries and Oceans. In addition, the Act directs the agency to have regard for the

sustainability of natural resources, assist in Canadian scientific and technological capabilities, help develop standards for the management of natural resources, participate in the marketing of the natural resources, and
gather and analyze data related to Canada's natural resources.[21]

Forestry Act 1989. This law established the Canada Forestry Service and
mandated a sustainable approach to forestry management.

Nuclear Energy Act 1985. This Act allows for research into and development of nuclear energy. While the federal government has important
responsibilities relating to nuclear energy, it primarily serves only in the
development of policy. The decision to invest in electric generation rests
with the provinces, and it is up to them to determine whether or not new
nuclear power plants should be built.

FISHERIES AND OCEANS CANADA

Fisheries and Oceans Canada   Pêches et Océans Canada   Canada

Department: Fisheries and Oceans Canada
Established in Current Form: 2008
Number of FTE Equivalents: ~ 10,000
Annual Budget: $C 1.7 billion
Activities: Contains the Canadian Coast Guard; oversees boating safety on
freshwater and marine areas; regulates harvest of fish, shellfish and other
marine organisms; regulates commercial harvest of freshwater fish;
produces and stores data on tides, currents and water levels; protects
marine mammals.
Important Laws or Acts: Canada Shipping Act 2001, Coastal Fisheries
Protection Act, Department of Fisheries and Oceans Act, Fisheries Act,
Oceans Act, others.

The Department of Fisheries and Oceans (DFO) is the lead federal
agency for maritime activities and commercial fishing operations. It contains
the Canada Coast Guard, which protects the shores of Canada, conducts boat
and ship inspections, regulates the commercial harvest and processing of fish
and other aquatic food organisms in both freshwater and oceans, oversees
aquacultural activities in the nation, regulates recreational fishing in saltwater habitats, provides maritime weather data and protects marine mammals,
including those that are considered at risk under the Species at Risk Act.

Given the range of activities administered by the DFO, the agency is
responsible for overseeing a sizeable portion of Canadian income (Table 5.1).
Hopefully your professor won't require you to memorize these numbers, but
they are impressive. In 2006, private commercial fishing, for example, had

TABLE 5.1 Economic Values of Activities tied to Fisheries and Oceans Canada

Activity	Revenues (C$ thousands)	Number of jobs*
Commercial saltwater harvest	1,820,281	40,000
Saltwater aquaculture	895,031	5,450
Saltwater processing	3,962,305	38,100
Freshwater commercial harvests	68, 670	500
Saltwater recreational	778,000	16,482
Cruise ship tourism	472,000	9,990
Coastal tourism	2,200,000	65,424
Offshore gas and oil extraction	9,289,130	8,417
Marine transport	6,270,000	77,960
Marine construction	419,679	5,610
Shipbuilding and repairs	1,065,000	16,066

*Estimate formed by the number of commercial licenses sold[24] in 2006 × an average employee number of 5 – this last number is a crude estimate.

21,000 licensed vessels, employed around 40,000 people, and harvested over 1 million tonnes (or 1.13 short tons) of raw product. This grossed C$1.8 billion.[22] In the same period, aquaculture resulted in 5450 jobs and 167,800 tonnes of seafood, and grossed C$895 million. Processing plants led to an additional 38,100 jobs. Altogether, the seafood industry amounted to over C$4.8 billion, 81% of which was due to exports, primarily to the United States. Freshwater harvests were smaller, but in 2007 Canadian commercial fishermen harvested 32,000 tonnes of freshwater fish. More than 26 species were harvested, but the largest catches were walleye (*Stizostedion vitreum)*, whitefish (*Coregonus* spp.) yellow perch (*Perca flavescens*), and northern pike (*Esox lucius*). This resulted in an additional C$69 million to the Canadian economy.[23] Commercial seafood, including freshwater and saltwater varieties, is the largest food export in Canada in terms of dollars.

Recreational activities are also lucrative. Fishing in estuaries, nearshore and offshore resulted in approximately 3.2 billion fisherman/days; these anglers spent an estimated C$778 million annually[19] and accounted for approximately 16,500 jobs. Cruise ship tourism grossed C$472 million and produced almost 10,000 positions. Coastal tourism, which includes auto tours, beach excursions and all the other activities that take place on shorelines and near shore water, was huge, earning C$2.2 billion and employing 65,424 workers, in restaurants, souvenir shops, beach rentals and so on.

Offshore oil and gas extraction in Canada yielded about C$9.2 billion in sales with 8417 jobs in 2006.[19] Marine transport grossed revenues of around C$6.3

billion with funds coming both from the direct fees and costs of water transportation and from support activities such as docking charges, port and harbor fees and cargo handling. Fisheries and Oceans Canada also has responsibility for marine construction and shipbuilding. While neither of these are natural resources *per se*, they are intimately tied to the ocean, which is a natural resource. Together, they produced C$1.5 billion in revenues and more than 21,000 jobs if both direct industrial and supporting services are included. Altogether, the DFO administers industries that gross over C$25 billion annually.

Organizational Structure of Fisheries and Oceans

From an organizational perspective, the DFO is represented in each of six regions by a regional director general (Figure 5.10). The Coast Guard is

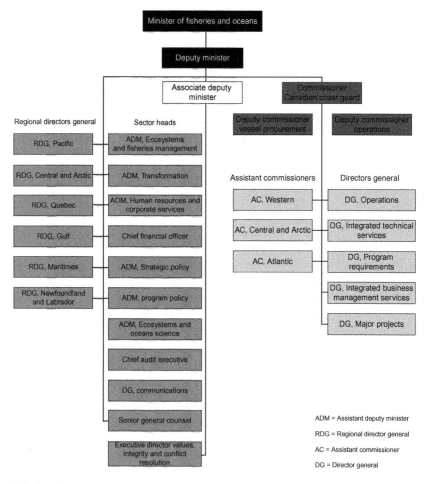

FIGURE 5.10 Organization structure for Fisheries and Oceans. *Credit: Fisheries and Oceans*

headed by a Commissioner who has line authority under the Deputy Minister but above the Associate Deputy Minister. The Coast Guard is further directed by two Deputy Commissioners, one for vessel procurement, which oversees the purchase and construction of new ships or boats for the ministry. The other Deputy Commissioner oversees the operations of the Coast Guard. In line with other natural resource ministries, there is a group of Sector Heads or Assistant Deputy Ministers (ADM) that work with conducting the business affairs of the ministry — Communications, Counsel, Values, Finances, Strategic Policy, and Audits. In addition, the *Ecosystems and Fisheries Sector* has two major groups of functions. One of these is to provide protection and management of the oceans around Canada. This protection extends to the development of marine sanctuaries, concern for species at risk, establishment of water quality standards, provision of guidance to landowners along the shoreline to avoid pollution and shoreline degradation, and public education about the marine environment.[25] The other group of functions is the management of all things dealing with commercial fisheries. This includes monitoring and regulating saltwater and freshwater harvests, licensing aquaculture facilities, monitoring saltwater recreation (freshwater recreational fishing is the responsibility of the provinces) and maintaining data on harvests. While Ecosystems and Fisheries conducts the management end of the agency, *Ecosystems and Oceans Science* is the science and research end. This sector monitors the ecological health of saltwater environments, explores for and monitors oil and gas deposits, operates three centers of expertise and conducts climate research.[26] As of this writing the ministry has been downsized and may experience additional budget cuts, so many functions have been combined into relatively few sectors to save money and allow more efficient operation.

Fisheries and Oceans Budget

Around 43% of the annual budget for the DFO is spent on Safe and Secure Waters. This includes the Coast Guard (Figure 5.11) and oversight on maritime construction and shipbuilding. Around one-quarter of the budget goes to the *Prosperous Maritime Sectors and Fisheries*; this is spread equally between direct conservation of fisheries and oceans, and financial returns from ocean and freshwater harvests. *Operations* accounts for about 17% of the total budget, and covers the costs of running the ministry.

Selected Acts under Fisheries and Oceans Administration

Fisheries and Oceans oversees 13 federal acts. Some of these acts relate to activities that are not tied directly to natural resources but affect shipping, boat construction and other marine industries. Below is a selected list of laws that provide further insight into the functions of the agency.

Percent

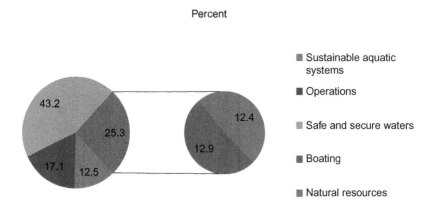

FIGURE 5.11   Annual budget for Fisheries and Oceans in terms of percent total budget. *Data: Treasury Board of Canada Secretariat. This figure is reproduced in color in the color plate section.*

Canada Shipping Act 2001. This Act replaced the Act of the same name that was passed in 1985. This is a broad act covering over 55 sets of regulations for oceanic shipping. While it is very important to the shipping industry and to the operation of the ministry, it is very limited in its application to natural resources — mostly regulations on bilge water discharges and pollution.

Coastal Fisheries Protection Act 1985. This law sounds like it may have something to do with natural resources, and it does — but in a very indirect way. It provides authority for Canada to license the entry and activity of foreign fishing vessels in Canadian waters and their access to Canadian ports. It can be a measure to regulate fish harvests.

Department of Fisheries and Oceans Act 1985. This Act establishes the powers, duties and functions of the Minister of Fisheries and Oceans Canada, including seacoast and inland fisheries; fishing and marine sciences; and the coordination of the policies and programs of the Government of Canada regarding oceans.

Fisheries Act 1982. This broad Act predates the Department of Fisheries and Oceans and goes back to the Confederation of Canada. It is the heart of the oceanic and commercial fishing of Canada. It applies to all fishing zones, territorial seas and inland waters of Canada, and is binding to federal, provincial and territorial governments. As federal legislation, the Fisheries Act supersedes provincial legislation when the two conflict. Consequently, approval under provincial legislation may not necessarily mean approval under the Fisheries Act.[27] The Constitution Act of 1982 further distinguished the roles of provincial and federal governments with regard to maritime functions, and gave the federal government authority over seas, coastal and inland fisheries, navigation, and fiduciary responsibility to aboriginal people. However, the government of Canada has essentially no control over the use of inland waters, watercourses or shorelines, which fall under provincial jurisdiction. Conversely, the provinces cannot make regulatory decisions concerning fish habitat. The Act covers many different areas dealing with fisheries, including commercial fishing regulations in each province, management of contaminants, marine mammals, fish health, metal mining effluents, pulp and paper effluents, and regulations in both the Pacific and Atlantic Oceans.

Oceans Act 1996. (amended 2005). This law establishes that the Minister of Fisheries and Oceans has authority over ocean management responsibilities and the Canada Coast Guard. It sets the management strategy that the ministers must follow, and identifies Canada's maritime zones. Specific regulations were also established for marine sanctuaries.

Study Questions

5.1 Describe the primary functions of Environment Canada, Fisheries and Oceans, and Natural Resources Canada.

5.2 To which ministries do the following departments belong? Canadian Wildlife Service, Parks Canada, Canadian Forestry Service.

5.3 In which major ways does the parliamentary system in Canada differ from the federal government structure of the United States? In which ways are they similar?

5.4 Describe and compare the relationship between the federal and provincial governments in Canada as they deal with natural resources. How does this relationship compare to that between states and the federal government of the United States?

REFERENCES

1. Makarenko J. The monarchy in Canada. Mapleleafweb. <www.Mapleleafweb.com>; 2007.
2. Makarenko J. Provincial Government in Canada: Organization, Institutions & Issues. Mapleleafweb. <www.mapleleafweb.com>; 2009.
3. Hogg P. *Constitutional Law of Canada.* 5th ed. Scarborough, Ontario: Carswell; 2006.
4. The Constitutional Distribution of Legislative Powers. Canadian Privy Council Office. <www.pco-bcp.gc.ca>.
5. Paehlke R. The environmental movement in Canada. In: VanNijnatten DL, Boardman R, eds. *Canadian Environmental Policy and Politics.* Don Mills, Ontario: Oxford University Press; 2009:2–13.
6. Toner G, Meadowcraft J. The struggle of the Canadian federal government to institutionalize sustainable development. In: VanNijnatten DL, Boardman R, eds. *Canadian Environmental Policy and Politics.* Don Mills, Ontario: Oxford University Press; 2009:77–90.
7. Harrison K. *Passing the Buck: Federalism and Canadian Environmental Policy.* Vancouver, BC: University of British Columbia Press; 1996.

8. Parliament of Canada. Canadian Ministry. <http://www.parl.gc.ca/MembersOfParliament/MainCabinetCompleteList.aspx?TimePeriod = Current>.

9. Environment Canada. About Environment Canada. <http://www.ec.gc.ca/default.asp?lang=En&n=BD3CE17D-1>; 2013.

10. Canadian Wildlife Service. <http://en.wikipedia.org/wiki/Canadian_wildlife_service>.

11. Treasury Board of Canada. <http://www.tbs-sct.gc.ca>.

12. Environment Canada. Acts. <http://www.ec.gc.ca/default.asp?lang = En&n = E826924C-1>; 2013.

13. Environment Canada. Natural Wildlife Areas. <http://www.ec.gc.ca/ap-pa/default.asp?lang=En&n=2BD71B33-1>; 2013.

14. Wikipedia. National Parks of Canada. <https://en.wikipedia.org/wiki/National_parks_of_Canada>; 2013.

15. Parks Canada. About Us. <http://www.pc.gc.ca/eng/agen/index.aspx>; 2013.

16. Treasury Board of Canada. <http://www.tbs-sct.gc.ca>.

17. Natural Resources Canada. NRCan: Firsts and Fascinating Facts from its Illustrious Past. <http://www.nrcan.gc.ca/history/562>; 2010.

18. Natural Resources Canada. The Department. <http://www.nrcan.gc.ca/department/535>; 2013.

19. Action for our Planet. The largest forests. <http://www.actionforourplanet.com/#/top-10-largest-forests/4559732818>; 2013.

20. National Resources Canada. About Us. <http://cfs.nrcan.gc.ca/about>; 2013.

21. Government of Canada. Justice Laws website. <http://laws-lois.justice.gc.ca/eng/acts/N-20.8/page-2.html#docCont>; 2013.

22. Fisheries and Oceans Canada. Economic impact of marine related activities in Canada. <http://www.dfo-mpo.gc.ca/ea-ae/cat1/no1-1/no1-1-econoprivate-eng.htm>; 2013.

23. Fisheries and Oceans Canada. Canadian Fishing Industry Overview. Economic Analysis and Statistics. <http://www.apcfnc.ca/en/fisheries/resources/Aboriginal%20Fisheries%20in%20Canada%20-%20Overview%20-Canadian%20Market%20Trends%20-%20David%20Millette.pdf>; 2011.

24. Fisheries and Oceans Canada. <http://www.dfo-mpo.gc.ca/stats/>.

25. Fisheries and Oceans Canada. <http://www.mar.dfo-mpo.gc.ca/e0010340>.

26. Fisheries and Oceans Canada. Ocean and ecosystems science. <http://www2.mar.dfo-mpo.gc.ca/science/ocean/sci/sci-e.html>; 2013.

27. Fisheries and Oceans Canada. Fisheries Act. <http://www.dfo-mpo.gc.ca/habitat/role/141/1415/14151-eng.htm>; 2013.

Chapter 6

US Department of the Interior

Terms to Know

- National Wildlife Refuge System
- Ecological Service Office
- Division of Economic Ornithology and Mammalogy
- Division of Biological Survey
- Pelican Island
- Jay Norwood "Ding" Darling
- Bureau of Fisheries
- US Fish and Wildlife Service
- National Wildlife Refuge Systems Improvement Act
- National Park Service
- Yellowstone National Park Act
- Antiquities Act
- Organic Act
- Historic Sites Act
- National Park Omnibus Management Act 1998
- Bureau of Land Management
- Land Ordinance of 1785
- Homestead Acts
- Taylor Grazing Act 1934
- United States Geological Survey
- Cooperative Research Units
- Bureau of Indian Affairs
- Bureau of Ocean Energy Management, Regulation and Enforcement
- Bureau of Safety and Environmental Enforcement
- Bureau of Reclamation
- Office of Surface Mining Reclamation and Enforcement

D.W. Sparling: Natural Resources Administration. DOI: http://dx.doi.org/10.1016/B978-0-12-404647-4.00006-4

AN OVERVIEW OF THE DEPARTMENT OF THE INTERIOR

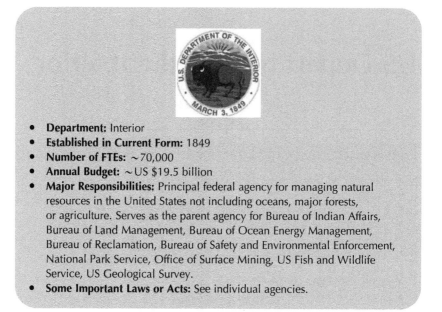

- **Department:** Interior
- **Established in Current Form:** 1849
- **Number of FTEs:** ~70,000
- **Annual Budget:** ~US $19.5 billion
- **Major Responsibilities:** Principal federal agency for managing natural resources in the United States not including oceans, major forests, or agriculture. Serves as the parent agency for Bureau of Indian Affairs, Bureau of Land Management, Bureau of Ocean Energy Management, Bureau of Reclamation, Bureau of Safety and Environmental Enforcement, National Park Service, Office of Surface Mining, US Fish and Wildlife Service, US Geological Survey.
- **Some Important Laws or Acts:** See individual agencies.

The *Department of the Interior* (DOI) is the principal federal department for managing natural resources other than agriculture, oceans and forests in the United States. Its mission statement, "The US Department of the Interior protects America's natural resources and heritage, honors our cultures and tribal communities, and supplies the energy to power our future",[1] reflects an agency with a great deal of responsibility and a multiplicity of functions. During the early decades of the United States, Congress was more concerned about the possibility of war with Great Britain, making treaties with other nations and finding money to keep the government going than with the natural resources of the country. Also, at that time the general sentiment was that the nation's resources and environment would never be exhausted, so why worry about preserving them? Eventually, in 1849, Congress passed an Act to create the DOI and gave it the responsibility of handling all of the internal affairs of the nation, including construction of the national capital's water system, colonization of freed slaves in Haiti, exploration of the western wilderness, oversight of the District of Columbia jail, regulation of territorial governments, management of hospitals and universities, management of public parks, and the basic responsibilities for Indians, public lands, patents, and pensions.[1] Note that natural resource management did not take precedence in this long list of responsibilities. Gradual loss and accretion of responsibilities over the decades, however, led to a Department whose primary concern is the wildlife, fish, natural areas, water, minerals, and energy resources of the country. Later in this chapter specific historical details will be presented concerning the agencies within the Department of the Interior.

Today, the Department employs nearly 70,000 people stationed in 2800 locations across the United States, Puerto Rico, and its territories. These employees' functions include scores of different job categories, ranging from biologists, geologists, hydrologists, meteorologists and climate specialists to social scientists, budget experts, secretaries and office staff.

Organization of the Department of the Interior

The organizational chart of the DOI helps us to understand the structure of this diversity (Figure 6.1).[1] Like all departments within the President's Cabinet, the DOI is headed by a secretary, who is supported by a deputy secretary and assistant secretaries. The Secretary and Deputy Secretary of the Department of the Interior are Presidential appointments requiring the confirmation of the United States Senate. The Assistant Secretary of the Interior for Fish, Wildlife and Parks, and directors of specific agencies within the DOI are also Presidential appointees. Support offices include the Solicitor and Inspector General, who provide legal advice and oversight; and the assistant secretary for policy, management budget and business that handles the everyday business affairs of the agency. In many ways federal agencies are similar to huge corporations, and need staff that take care of running the operations – making sure that employment practices meet federal guidelines, managing the preparation of the budget and then executing it through accounting practices, purchasing, contracting, personnel management, and so forth. Other offices pay attention to real estate, information technology, and both internal and external communications. Subgroups with the DOI, such as the US Fish and Wildlife Service or National Park Service, have corresponding offices to serve their level of government, but Headquarters for the Department oversees these. The other assistant secretaries supervise the offices and functions that actually deal with natural resources. Note that specific details of any agency organization chart are subject to change as new secretaries or agency directors take office, or as national priorities shift. Thus the organizational charts that appear in this book may be somewhat different from those that exist at the time you read it, so it is probably more important to understand something about organizing agencies rather than memorizing specific organizational charts.

Department of the Interior's Budget

Another factor that definitely varies from year to year is the actual amount of money budgeted to a department. Each department in the Cabinet vies

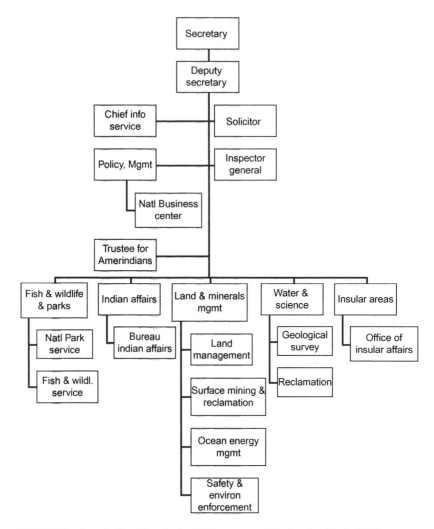

FIGURE 6.1 Organization Chart for the US Department of the Interior. *Credit: US Department of the Interior.*

with all of the other departments for their share of the money pot. As national priorities change, so too do budget allocations. If there is an ongoing war or major military action, for example, the Department of Defense may receive a larger proportion of the overall budget than if the nation is at peace. As mentioned in Chapter 12, there are really several budgets. The process starts with the President of the United States, but it continues to the

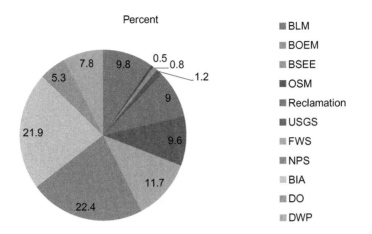

Percent

- BLM
- BOEM
- BSEE
- OSM
- Reclamation
- USGS
- FWS
- NPS
- BIA
- DO
- DWP

FIGURE 6.2   Annual Budget for the Department of Interior by Agency. *Credit: US Department of the Interior. This figure is reproduced in color in the color plate section.*

House of Representatives and Senate, and finally to a compromise or enacted budget that all three groups sign. For example, for the 2012 budget the President requested over $17.8 billion for the Department of the Interior, but the final approved budget was $19.46 billion; the difference was due to Congressional add-ons. Administrators within an agency like to see the budget increase because fixed costs tend to increase due to inflation and other factors, but in difficult economic times budgets may decrease. A 10% budget cut from the previous year would be considered severe. We will generally present enacted budgetary figures as percent of total agency budgets rather than as actual dollars, because there is less variation in percentages − a program or agency that receives a large portion of a budget will, in general, continue to receive a large portion from year to year; similarly, small segments tend to remain that way.

The largest agencies in terms of dollars within the DOI are the Bureau of Indian Affairs (BIA) and the National Park Service (NPS) (Figure 6.2). The Fish and Wildlife Service (FWS), Geological Survey (USGS), Reclamation, and Bureau of Land Management (BLM) are intermediately funded. The Office of Surface Mining (OSM), Bureau of Ocean Energy Management (BOEM) and Bureau of Safety and Environmental Enforcement (BSEE), all of which are more regulatory than managerial, are the smallest agencies. The Departmental Operations (DO) and department-wide programs (DWP) are mostly involved with the internal activities of the department.

UNITED STATES FISH AND WILDLIFE SERVICE

- **Agency:** US Fish and Wildlife Service
- **Parent Department:** Department of Interior
- **Established in Current Form:** 1974, 1993
- **Number of FTEs:** 9100–9600
- **Annual Budget:** $1.4–1.7 billion
- **Responsibilities:** National Wildlife Refuges, Ecological Services Offices, Federal Endangered Species, Anadramous fishes, International Agreements, Federal Conservation Law Enforcement, Migratory Bird.
- **Some Important Laws or Acts:** Lacey Act 1900, Migratory Bird Conservation Act 1929, Migratory Bird Hunting Stamp Act 1934, Federal Aid in Wildlife Restoration Act 1937, Endangered Species Act 1973, Alaska National Interest Lands Conservation Act 1980, National Wildlife Refuge System Improvement Act 1997.

The *United States Fish and Wildlife Service* (FWS) is the principal federal agency responsible for fish and wildlife management in the United States. Among other services, the FWS manages the *National Wildlife Refuge System* and *Ecological Services Offices* across the country. It has primary responsibility for enforcing the *Endangered Species Act*, the *Migratory Bird Treaty Act*, the *Lacey Act, Federal Aid to Fish and Wildlife Acts* and others. It is also involved in international wildlife laws and trade, such as the Convention on International Trade in Endangered Species of Wild Flora and Fauna (CITES).

According to their official website (in 2013)[2]: "The US Fish and Wildlife Service's mission is working with others, to conserve, protect and enhance fish, wildlife, and plants and their habitats for the continuing benefit of the American people."

History of the Agency

The precursors to the FWS were two other agencies. In 1871, the US Congress created the *Commission on Fish and Fisheries* within the Department of Commerce to study the declining food fishes in the United States, make recommendations on how to stop the decline, and establish aquaculture. Fourteen years later, the *Division of Economic Ornithology and Mammalogy*

FIGURE 6.3   Pelican Island, Florida – The first National Wildlife Refuge. *Credit: US Fish and Wildlife Service.*

was established by Congress in the Department of Agriculture. This agency's early work focused on controlling nuisance species of birds and mammals, primarily those that depredated crops (hence the "Economic" part of the title). This made sense, because it was located within the Department of Agriculture and crop protection was of greater concern than wildlife conservation at that time. However, the agency was also called to begin a survey to determine the geographic distributions of North American animals. Recall from Chapter 2 that in the late 1800s agencies were primarily concerned with controlling nuisance species and preserving natural resources, and a conservation attitude had not yet matured. In 1896, as concern for dwindling wildlife continued to grow, the name and mission of the agency were changed to the *Division of Biological Survey* and an increased focus on identifying the distribution and abundance of birds and mammals in the nation.

In the early 20th century, two events enhanced the wildlife element of the future FWS. In 1900, the *Lacey Act* was passed by Congress (see Chapter 3). This was the first federal wildlife law. The Division of Biological Survey was designated as one of the principal enforcement agencies for the Act. Since 1900, the Lacey Act has been applied countless times to felonies dealing with illegal wildlife and plant trade.

The other development of the early 1900s was the establishment of *Pelican Island*, Florida, as the first federal bird sanctuary (Figure 6.3). Pelican Island is a 5-acre (2 ha) mangrove island heavily populated by brown pelicans (*Pelecanus occidentalis*) and shorebirds. At the turn of the century the birds were threatened by the millinery industry, which used feather plumes for ladies hats and other fashions, and there were no federal or state laws to protect them. The naturalist and then curator of the American Museum of Natural

History in New York, *Frank Chapman*, recognized that the island was the last brown pelican rookery remaining on the east coast of Florida. Chapman, with the American Ornithologist's Union and Florida Audubon Society, encouraged the state of Florida to pass a law protecting non-game birds. Further protection was gained when President Theodore Roosevelt, a noted conservationist, signed an Executive Order to establish Pelican Island as a federal bird reservation under the Division of Biological Survey. During his term of office Roosevelt set aside 54 other reservations which would eventually become the seeds of the *National Wildlife Refuge* system, but Pelican Island was the first.

With these and other added duties, the Division was renamed the *Bureau of Biological Survey* in 1905 — a name that persisted for 34 years and through many changes in duties. During this time the *Migratory Bird Treaty Act* was enacted in 1918, with the cooperation of Great Britain (for Canada) and the United States. Later, conventions were added with Mexico, Japan and Russia to provide further protection of migratory birds. Chief US responsibility for this Act was given to the Bureau. During the Great Depression of the 1930s President Franklin D. Roosevelt established the Civilian Conservation Corps, which employed thousands of men and women to construct and bolster the infrastructure of over 50 National Wildlife Refuges.

As the desire to add new refuges to the system grew, a source of funding was necessary to purchase desirable lands. In 1934 Congress provided a means of obtaining funding through the *Migratory Bird Hunting Act*, otherwise known as the *Duck Stamp Act*. To hunt waterfowl legally, sportsmen now needed to purchase a stamp which they affixed to their hunting licenses. The first duck stamp cost US$1. When adjusted for inflation over all those years, the stamp was actually more expensive than the $15 it costs now. A few months before the Duck Stamp Act became law, one of the most imaginative and enthusiastic men in the early conservation movement was appointed Chief of the Bureau of Biological Survey. *Jay Norwood "Ding" Darling* had a sincere interest in nature and the environment since he was a child. While in college he found that he had a real talent for satirical or political cartoons and he became famous for taking corporations to task (Figure 6.4). Early in his career Darling became good friends with Franklin Delano Roosevelt, and later was appointed temporary Chief of the Bureau. Darling infused new energy into the bureau and obtained increased funding; he was instrumental in getting the Duck Stamp passed, and even designed the first stamp.

By now we were well into the period of protectionism and getting into the era of scientific management of wildlife (see Chapter 2). In 1937 the *Federal Aid in Wildlife Restoration Act*, also known as the *Pittman-Robertson Act*, was enacted. This Act has done much to stimulate wildlife management and research at the state level (see Chapter 12).

Recognizing that fisheries and wildlife had a lot in common, the *Bureau of Fisheries* was moved into the Department of the Interior in 1939, and the

FIGURE 6.4 Political cartoon by Jay "Ding" Darling, a leading environmentalist in the 1930s. *Reproduced with permission of the Jay Darling Foundation.*

Bureau of Biological Survey was transferred from the Department of Agriculture. A year later they were united as one agency: the first *Fish and Wildlife Service*. The current flyway system, consisting of four flyways, was established in 1947, and funding for fisheries was promulgated by the *Federal Aid in Sport Fish Restoration Act* (also known as *Dingell-Johnson*) in 1950 (see Chapter 12).

Conflicts of interest existed in the early Fish and Wildlife Service. There were some administrators and biologists who favored sport hunting and recreational fishing and shunned commercial fishing. Others were more interested in commercial fisheries. Thus, in 1956 the first Fish and Wildlife Service was split into the *Bureau of Commercial Fisheries* and the *Bureau of Sport Fisheries and Wildlife*. The Bureau of Commercial Fisheries eventually became part of the Department of Commerce as the *National Marine Fisheries Service*, and, with those commercial concerns gone, the Bureau of Sport Fisheries and Wildlife was renamed as the second *Fish and Wildlife Service* in 1974.

In 1973 the Bureau or Sport Fisheries and Wildlife (later FWS) received a major responsibility when the *Endangered Species Act* was passed (more on this Act can be found in Chapter 4). The FWS assumed primary enforcement of the Act and all its ramifications for all species except for some marine mammals. Managing the Endangered Species program has become one of the major functions of the FWS. In 1980, the size of the National Wildlife Refuge nearly doubled in size with the passage of the *Alaska National Interest Lands Conservation Act*. The Act added 16 new refuges and 77 million acres (31 million ha) to the system. Altogether, 100 million acres (40.47 million ha) of natural lands in Alaska were protected with one stroke of the President's pen.

Another conflict of interests arose in the 1980s, similar to that of the 1950s. A portion of the scientists in FWS were conducting studies on reducing wildlife depredation on crops and other nuisances while a majority of biologists within the agency were interested in conservation, not control. As a result, the *Animal Damage Control* section of FWS was moved to the Department of Agriculture under the Animal and Plant Inspection Service (APHIS), and formed the Denver Wildlife Research Center. A second big shift in the FWS and other DOI agencies occurred in 1993, when all research in the Department of the Interior was moved to the US Geological Survey – but we'll deal with that in greater detail later in this chapter.

Recall that the first National Wildlife Refuge, Pelican Island, was established by an Executive Order. That meant that Congress never officially approved the formation of the National Wildlife Refuge System. In 1997, some 96 years later, that was rectified when the *National Wildlife Refuge Systems Improvement Act* was signed. This served as the "organic" Act for the refuge system, in that it included Congressional approval, and direction as to the mission of the refuges.

Organizational Structure of the Fish and Wildlife Service in Washington, DC

A recent organizational flow chart for the Fish and Wildlife Service is shown in Figure 6.5.[3] Recall that this chart is for only one part of the Department of the Interior, which is only one part of the Executive Branch of the federal government. This is what is meant as government bureaucracy! Rather than go into excruciating detail of the agency, some of which could change, I'll touch upon some of the highlights.

It's apparent that there are many parallels between the organizational chart for FWS and that for DOI. The organizational chart is top-down, with various associate department heads reporting to deputy assistant directors – assistant directors – deputy directors and, finally, the director, who reports to the Secretary of the Interior, who reports to the President of the United States. All the levels above regional directors in this flow chart are

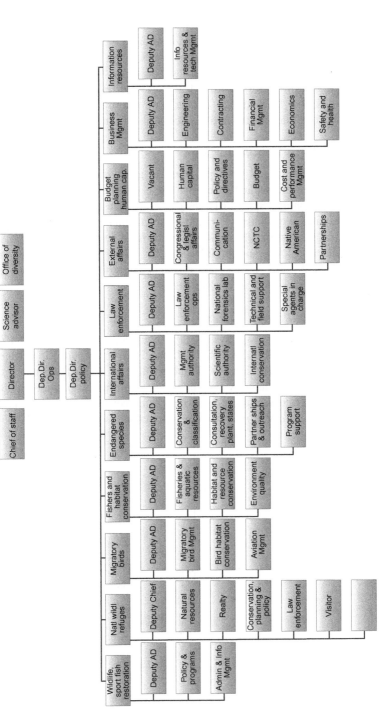

FIGURE 6.5 Flow chart for the US Fish and Wildlife Service. *Credit: US Fish and Wildlife Service.*

headquartered in or near Washington DC. In addition to operations or business divisions, there are divisions that deal more directly with the natural resources entrusted to the agency. In the FWS, these natural resource (or "hands on") divisions include Wildlife and Sport Fish Restoration, National Wildlife Refuges, Migratory Birds, Fisheries and Habitat Conservation, Endangered Species, International Affairs, Law Enforcement, and External Affairs. Below is a brief description of each of these functional offices.

Wildlife and Sport Fish Restoration

This office manages several grant programs to assist states in their wildlife and fisheries management and research. These include the Federal Aid in Wildlife Restoration and Sport Fisheries programs. These cost-sharing grants go to safe boating and hunting programs, habitat improvement and acquisition, and research. Historically, these programs have been closely associated with game species of fish and wildlife. Smaller grant programs, such as the State and Tribal Wildlife Grants, are also under this office, and may cover species other than those that are harvestable.

National Wildlife Refuge System

The National Wildlife Refuge System is a dynamic, growing collection of protected areas for wildlife. As of 2012, the FWS managed 556 refuges encompassing 150 million acres (60.7 million ha) and 38 wetland management districts. Most of the early refuges were established to provide habitat and shelter for migratory birds, so they tend to be concentrated along major flyways (Figure 6.6).[4] A large cluster of refuges and waterfowl production

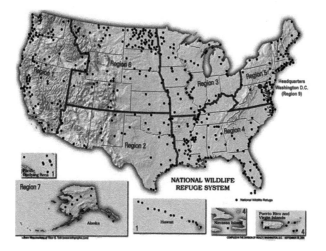

FIGURE 6.6 Location of the National Wildlife Refuges, US Fish and Wildlife Service. *Credit: US Geological Survey.*

areas also exists in the northern prairie states of North Dakota, Minnesota, Montana and South Dakota, which are major breeding areas for waterfowl. Wildlife refuges are home to more than 700 species of birds, 220 species of mammals, 250 reptile and amphibian species, and 200 species of fish.[5] In addition to migratory birds, some refuges have been established to protect threatened or endangered species. For example, the Sauta Cave and Fern Cave refuges in Alabama are relatively small (<300-acre [121 ha]) sites for the endangered Indiana bat (*Myotis sodalis*) and gray bat (*M. grisescens*). The Watercress Darter NWR is a 7-acre (1.7 ha) site established specifically to protect the endangered watercress darter (*Etheostoma swaini*), a small fish (Figure 6.7). Wildlife are not the only organisms protected by national wildlife refuges; the San Diego NWR, California, was established to ensure habitat for four species of plants and a fairy shrimp.

Some National Wildlife Refuges are very isolated, occurring hundreds of miles from any form of urban area, and receive few visitors. For example, Isembek NWR is on the Aleutian Peninsula, more than 600 miles (966 km) from Anchorage, with no road access and only a few visitors each year. In contrast, Crab Orchard NWR in southern Illinois is the most visited of all refuges, with over 1 million visitors per year.

The refuge system was established with one main purpose: to provide habitat for fish and wildlife. Through the decades, however, there has been considerable pressure from the public to widen the recreational opportunities within refuges. Many of the proposed activities could be detrimental to the wildlife inhabiting these refuges, and are rejected. With the 1997

FIGURE 6.7   Watercress darter (*Etheostoma swaini*), an endangered species that has its own National Wildlife Refuge. *Credit: Outdooralabama.com*

National Wildlife Refuge Systems Improvement Act, Congress established six primary wildlife-related activities that refuges can provide, if compatible with refuge operations. These include hunting, fishing, wildlife observation, photography, interpretation, and education[5]. However, some refuges offer more than those six primary uses. Crab Orchard NWR, for example, was established soon after World War II and its property included munitions plants and bunkers that were still in use, so industry was declared a compatible use for that refuge. Another unusual mandate for Crab Orchard was to provide opportunities for farming, and the refuge leases lands for share-cropping, some of which in turn provides winter foods for waterfowl and deer. Since the late 1980s, the Department of Defense has been closing or reducing the size of many military installations around the country. Much of the released lands went to nearby refuges, with some active operations continuing. Big Oaks NWR in southeastern Indiana, for example, was created from lands obtained from the Jefferson Proving Grounds, and shares its property with the Indiana National Guard, which practices aerial maneuvers on parts of the refuge. Patuxent Wildlife Research Refuge is unique in the system because it was established as a research refuge when research was still a part of the FWS. The refuge currently houses the US Geological Survey's Patuxent Wildlife Research Center. The research center uses the refuge for studies as part of an agreement between the two agencies.

Migratory Birds

The Migratory Bird Treaty Act 1918 gave the FWS (then Bureau of Biological Survey) principal responsibility for conserving migratory birds within the United States. By definition, a migratory bird is one that regularly crosses national borders as, for example, between breeding and wintering grounds.[6] Additional details of this Act can be found in Chapter 4.

The mission of the Fish and Wildlife Service's Migratory Bird Program "is to conserve migratory bird populations and their habitats for future generations, through careful monitoring, effective management, and by supporting national and international partnerships that conserve habitat for migratory birds and other wildlife".[6] The office works with other federal agencies, states and nations in conserving migratory birds. Some of these, such as ducks, geese, swans, doves, woodcock, rails and coots, are game birds that have received substantial attention over the years. Most are songbirds, such as Neotropical migrants, and many of these have been experiencing serious population declines over the past several decades. Activities of the office include monitoring migrant bird populations, enforcing laws, education, and issuing permits such as duck stamps and scientific research or collection permits.

Fisheries and Habitat Conservation

This office manages most of the FWS activities dealing with fish management and conservation. It includes Fish and Wildlife Conservation Offices, National Fish Hatcheries (Figure 6.8), Fish Health Centers, Fish Technology Centers, and the Aquatic Animal Drug Approval Partnership Program Office.[7] The office has been involved with fisheries conservation since 1871, when its roots were formed along with the Commission on Fish and Fisheries. Initially, the Commission and subsequent FWS fisheries offices were concerned with developing methods to culture fish for human consumption and supplement the dwindling native populations. Later, emphasis was placed on anadramous fishes — those that live primarily in the ocean but enter freshwater streams to breed, such as salmon and trout. Since the passing of the Endangered Species Act, the hatchery program has also been involved with culturing endangered species of fishes. There are 70 National Fish Hatcheries in 35 states, with a slightly disproportionate representation in Washington State. Over 100 different species of fish have been raised in the program.

The Fish Health Centers study fish diseases and parasites to provide assistance to hatcheries and biologists working with natural fish populations. Fish Technology Centers provide technical assistance to the hatcheries, and aquatic resource conservation. They study conservation genetics for fish, investigate optimal population densities for fish culture and nutritional needs, and are involved in other related activities. Biologists with the Aquatic Animal Drug Approval Partnership Program work towards obtaining FDA-approved and EPA-compliant new animal drugs for use in federal, state, tribal and private aquaculture programs throughout the United States.

FIGURE 6.8 Neosho National Fish Hatchery, Neosho MO — the oldest existing National Fish Hatchery. *Credit: US Fish and Wildlife Service.*

Endangered Species

Passage of the Endangered Species Act in 1973 gave the FWS primary responsibility for threatened and endangered species of plants and animals. A more detailed description of this powerful Act can be found in Chapter 4.

International Affairs

Due to a great many cooperative agreements, conventions and treaties with other nations, the FWS has the need for a separate office to manage these agreements and interact with other nations on matters of endangered species, importation and exportation of animals and plants and their products, and similar duties. The FWS also sends expert consultants and financial assistance to other nations around the world to help them in their conservation efforts. Poaching for bush meat, rhinoceros horn, pelts and skins, ivory, the pet trade and other products inflicts heavy tolls on wildlife around the world, and the office of International Affairs works closely with Law Enforcement and Endangered Species to curtail such illegal activities. The office also represents the United States on CITES issues.

Law Enforcement

Did you ever shoot a bird with a BB or pellet gun when you were a child? If so, you probably unwittingly committed a federal crime. Such crimes are seldom reported, and, when they are, justice would dictate that a child should be given a stern lecture by a Conservation Officer and let off. However, many much more serious crimes involving wildlife and endangered species occur often, and it's up to the FWS law enforcement officers to uphold national laws. Conservation Officers work at refuge and regional levels in the field and at airports and sea ports to inspect shipments entering or leaving the United States. They have the authority of any other federal law enforcement agent. Some of their specific activities involve breaking up international and domestic smuggling rings that target imperiled animals; protecting wildlife from environmental hazards and safeguarding critical habitat for endangered species; enforcing federal migratory game bird hunting regulations and working with states to protect other game species from illegal take, and to preserve legitimate hunting opportunities; working with international counterparts to combat illegal trafficking in protected species; training other federal, state, tribal, and foreign law enforcement officers; using forensic science to analyze evidence and solve wildlife crimes; and distributing information and outreach materials to increase public understanding of wildlife conservation and promote compliance with wildlife protection laws.[8]

External Affairs

The FWS often interacts with other agencies within the United States, including Congress, other federal agencies, states, non-government agencies

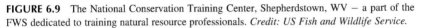

FIGURE 6.9   The National Conservation Training Center, Shepherdstown, WV – a part of the FWS dedicated to training natural resource professionals. *Credit: US Fish and Wildlife Service.*

and the American public. These activities are coordinated by the office of External Affairs. This office manages the National Conservation Training Center in Shepherdstown, WV, which provides training courses for staff of the FWS and other federal and state conservation agencies (Figure 6.9). Whenever Congressmen want information about natural resources in their districts or states, the office of External Affairs can expect a phone call. External Affairs is often the clearing house for media contacts and news releases. Staff within this office, while having good communication skills, must also be well informed about what is happening within the agency.

Organization Outside of Washington DC

While primary policy decisions are made at the Washington DC headquarters, implementation and often interpretation of that policy often occurs at the regional level. The FWS has eight regions (Figure 6.10) covering all of the United States, including Alaska, Hawaii, Puerto Rico, and territories. For the most part, regional offices have local counterparts to the divisions within headquarters. These offices serve as liaisons between Washington and the field. Actions at the regional office level often involve how best to adapt higher-level policy for implementation, to evaluate and report the effects of implementation, and to provide regional budgetary needs. In this way, regional offices provide line supervision over endangered species, fisheries, refuges, migratory birds, law enforcement, external affairs, and habitat in their part of the country. International interactions occur through the Washington office, but staff from the various regions may be included as necessary.

Fish and Wildlife Service Budget

As with any other federal agency, the budget for the Fish and Wildlife Service changes annually. During the first decade of this century, the total

FIGURE 6.10   Regions of the US Fish and Wildlife Service. *Credit: US Geological Survey.*

budget for the Service hovered around $1.4–1.7 billion. About 75% of the FWS budget can be described as discretionary and the remainder as fixed costs (Figure 6.11A). For FWS, fixed costs include outstanding indebtedness on land purchases and continuing commitments in support of individual state needs; these must be met. Discretionary funds pay for new and existing programs within the Service, are awarded (or not awarded) each year, and make their way through the budgetary process. Salaries within these programs are often considered fixed, but the truth is that they are included in the annual funding; if programs are slashed, staff may be transferred to other programs if this is possible.

After the fixed programs are paid for, the Fish and Wildlife Service distributes funding to eight compartments – the primary five divisions of the agency; ecological services that support Service programs at the state or regional levels; general operations or infrastructure support; and cooperative conservation which further assists states and provides grants for wildlife or habitat conservations (Figure 6.11B). Of the various compartments, Ecological Services and Refuges combined encompass more than 60% of the annual budget. Some of the funding in each compartment supports the national office, but most is distributed to the regional offices, which then partition their funding to respective offices.

As might be guessed, the primary financial compartment within National Wildlife Refuges is wildlife and habitat management. General maintenance – repairs, cleaning the buildings, snow removal, trail maintenance, and the like – is the next major compartment, followed by visitor services, which include hunting and fishing programs, and salaries for the people with warm, smiling faces that guide nature walks and related functions. Ideally, law

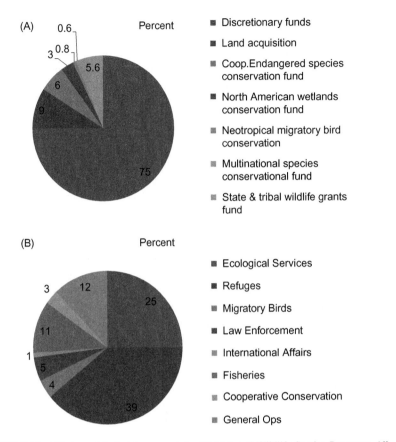

FIGURE 6.11   (A) Annual budget by percent for US Fish and Wildlife Service Programs. All but the discretionary funds portion are obligated even before the budget is made. (B) US Fish and Wildlife Service distribution of discretionary funds by percent. *Credit: US Fish and Wildlife Service. This figure is reproduced in color in the color plate section.*

enforcement would never be needed in a refuge, but that's just not realistic. Finally, planning entails activities such as holding public meetings, printing costs for reports, and staff time spent in developing comprehensive conservation plans and evaluating these plans on an annual basis.

Ecological Services has a comparatively simple budget consisting of only three categories: endangered species, habitat conservation, and environmental contaminants. As mentioned, Ecological Services offices are distributed throughout the country and are staffed with specialists in each of these categories to provide support to refuges and to states.

Each year the Fish and Wildlife Service surveys waterfowl breeding grounds in the Prairie Pothole region of Minnesota, North Dakota and South Dakota, where most of the ducks and geese in the United States are produced. They also monitor migratory songbird populations and conduct limited, non-refuge related habitat restorations; these functions consume most of the Conservation and Monitoring portion of the Migratory Bird division. The *North American Waterfowl Plan* was established in 1986 as an agreement between Canada and the United States to protect, enhance and restore habitat necessary for waterfowl. Various amendments since then have included Mexico as a national partner and developed localized joint ventures among federal and state agencies, non-government organizations, universities and volunteers to assist in habitat improvement. Avian Health and Disease funding supports research and mitigation of epizootics; the permit office issues scientific collecting and other permits dealing with migratory birds; and duck stamp offices manage the annual sale of duck stamps and the art competition associated with those stamps.

The chief expense of the fisheries program is to support the fish hatcheries across the country. The program also works towards assessing and improving stream and lake habitats and assessing the health of fish populations, especially those that spawn in fresh water but live most of their lives in oceans. Invasive and exotic species such as Asian carp, zebra mussels (*Dreissena polymorpha*), nutria (*Nutria canadensis*), sea lamprey (*Petromyzon marinus*), hydrilla (*Hydrilla verticillata*), Eurasian milfoil (*Myriophyllum spicatum*), purple loosestrife (*Lythrum salicaria*) and many others pose substantial problems to ecosystem health and are expensive to control, let alone eradicate.

NATIONAL PARK SERVICE

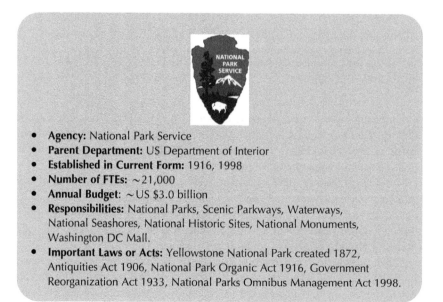

- **Agency:** National Park Service
- **Parent Department:** US Department of Interior
- **Established in Current Form:** 1916, 1998
- **Number of FTEs:** ~21,000
- **Annual Budget:** ~US $3.0 billion
- **Responsibilities:** National Parks, Scenic Parkways, Waterways, National Seashores, National Historic Sites, National Monuments, Washington DC Mall.
- **Important Laws or Acts:** Yellowstone National Park created 1872, Antiquities Act 1906, National Park Organic Act 1916, Government Reorganization Act 1933, National Parks Omnibus Management Act 1998.

If you have ever visited Yellowstone National Park in Wyoming, you undoubtedly marveled at the natural beauty of the evergreen forests and meadows, thrilled at the steam pools and geysers, and enjoyed the array of wildlife. Even with hundreds of other visitors around, these natural wonders amaze everyone who beholds them. Now imagine that you were one of the very first non-native Americans to witness these wonders, back in the very early 1900s. There are no other visitors, no crowds, no concessions stands — only you and a few of your most valued friends experiencing these amazing sites firsthand. Such were the conditions that President Theodore Roosevelt, John Muir and a few others witnessed in Yellowstone in 1903 (Figure 6.12). This is the type of wonder and awe that the National Park Service still tries to instill, often successfully, in its millions of visitors each year.

The mission of the National Park Service is to "preserve[s] unimpaired the natural and cultural resources and values of the national park system for the enjoyment, education, and inspiration of this and future generations".[9] The keywords here are "preserve unimpaired", not conserve. The National Park Service tries to maintain the natural beauty and functions of its parks with as little interference as possible. Active management or manipulation of habitat and landscapes is held to a minimum. In addition to the parks, most of which are west of the Mississippi River, the National Park Service also maintains national historical areas such as Gettysburg, national monuments

FIGURE 6.12   Theodore Roosevelt (left) and John Muir on a tour of Yellowstone area, 1903.

such as the Jefferson monument in Washington DC, scenic parkways such as the Baltimore–Washington Parkway, and scenic lake shores and seashores like Cape Cod; most of these occur in the eastern United States. Collectively, the NPS oversees 384 sites and over 83.6 million acres (33.8 million ha). Each year, over 280 million visitors come to the National Parks. Which one is visited most? That would be Great Smoky Mountain National Park (around 9 million visitors), due to its closeness to population centers and its beauty. Which gets the fewest number of visitors? That would probably be Kobuk National Park in Alaska near the Arctic Circle. It gets fewer than 1000 visitors a year, probably because the only way to get there is through walking, dog sled or snowmobile, and it has no designated roads or trails.

History of the Agency

The precursors to the National Park Service occurred in 1872 and 1906. In 1872, Congress set aside more than 2 million acres (0.81 million ha) in the Montana and Wyoming territories under the *Yellowstone National Park Act* to be "dedicated and set apart as a public park or pleasuring-ground for the benefit and enjoyment of the people".[10] Congress placed this park under the jurisdiction of the Department of the Interior to preserve all natural resources within the park. A few years before this, the federal government had purchased the lands that would make up Yosemite National Park and donated them to California as a state park. However, Yellowstone set the precedent for national parks, and by the early 1900s Sequoia, Mount Rainer, Crater Lake and Glacier had been added to the National Park gems, while California relinquished Yosemite to the Department of Interior. In 1906, the *Antiquities Act* was passed to preserve western historical treasures such as cliff dwellings and other Native American dwellings, early missions, and "other objects of historic or scientific interest" on federal lands. By 1916 the Department held 14 national parks and 21 national monuments and many historical sites, but still had no central coordination of these parks.

At this point Stephen Mather entered the cause for a centralized organization of parks. He was a persuasive, wealthy and well-connected businessman whose arguments gained him the position of Assistant Director of the Interior in Charge of Parks. Using his connections and clout, Mather teamed with his aide, Horace M. Albright, to convince Congress that a separate bureau needed to be created to manage the growing park system. In 1916, Congress passed the National Park Service organic Act or *Act to Establish the National Park Service*. Because this was an organic Act, Congress defined the responsibilities and mission of the NPS. Crucial to their mission was the "conservation of the scenery, natural and historic objects, and the wildlife therein and to provide for the enjoyment of the same in such manner and by such means as will leave them unimpaired for the enjoyment of future generations".[11]

The Secretary of the Interior directed Mather to place preservation as the primary goal for the NPS while simultaneously allowing people to visit and enjoy the sites. National Historic sites were to be maintained but not substantially altered. For parks, preservation suggested letting nature take its course while encouraging hotels, museums and interpretation for the benefit of visitors. For example, the NPS had a "natural burn" policy for many years. Under this policy, fires that were started by natural means such as lightening strikes were allowed to burn unless they threatened people or structures. Fires that were caused by campfires or carelessly tossed cigarettes, however, were extinguished as soon as possible. Unfortunately, a combination of "perfect storm" conditions in 1988 resulted in fires that eventually consumed 36% of Yellowstone National Park and a total of 1.4 million acres (0.57 million ha) in and around the park. This event resulted in a more stringent fire management policy within the Park Service. Nevertheless, direct management activities and habitat alteration in national parks are minimal compared to national wildlife refuges or national forests.

During the late 1920s and 1930s, the NPS expanded greatly. The new director, Horace Albright, persuaded President Franklin Roosevelt to make a massive reorganization in the Executive Branch of the federal government in 1933 and transfer all of the War Department's (now Department of Defense) monuments and parks to the Service. Similarly, all of the monuments in the Forest Service and those in Washington DC were placed under the National Park Service. Two years later the *Historic Sites Act* granted NPS substantial powers to conduct surveys of historic properties, acquire new sites, conduct research, identify historic sites under the management of other groups, and establish museums and interpretative activities for public education (Figure 6.13). For the next two years (1935 to 1937), the NPS continued to develop its holdings and responsibilities.

FIGURE 6.13 The Springfield home of Abraham Lincoln, one of the many historic sites managed by the National Park Service. *Credit: US National Park Service.*

Other major events occurred during the 1960s. In preparation for the celebration of the 50th anniversary of the National Park Service, Conrad Wirth, then director of the Service, instituted a major program to upgrade facilities and establish visitor centers in many parks and training centers in Arizona and West Virginia. A series of Congressional Acts during this time substantially augmented NPS duties. Throughout this period, however, maintaining natural beauty was strongly emphasized. The Leopold Report (1963) advised the Service that "the biotic associations within each park be maintained, or where necessary recreated, as nearly as possible in the condition that prevailed when the area was first visited by the white man ...". The *Wilderness Act* of 1964 called on the NPS to identify all wilderness areas on its lands and provide legal protection for such sites. The *Land and Water Conservation Fund Act* of 1965 provided funds for the expansion and acquisition of new parks to the system. In 1966 Congress passed the *National Historic Preservation Act*, which required all historical parks to enter into the National Registry of Historic Places. In addition, federal activities on these sites were subject to review by state historic preservation offices; this gave states some local input into the process of identifying sites of historic significance. The NPS was one of several federal agencies to assume responsibility for preserving stretches of rivers under the *Wild and Scenic Rivers Act* in 1968. As of 2011, there were almost 12,600 miles (21,000 km) of riverways on over 200 rivers in the system.[12] National scenic trails under the auspices of the NPS and US Forest Service were designated under the *National Trails System Act*, 1968; both the Appalachian Scenic Trail from Georgia to Maine and the Pacific Crest National Scenic Trail from California to Washington were included in this Act.

Recall that in 1980 Congress enacted the Alaska National Interest Lands Conservation Act, which greatly added to the National Wildlife Refuge System. This Act also added more than 47 million acres (19 million ha) to the park system. This was in addition to 45 million acres (18.2 million ha) of national monuments that had been added two years earlier. The 1980 legislation converted most of these monuments to national parks. As of 2012, there were 23 national parks in Alaska alone (Figure 6.14).

Prior to 1993, NPS had its own researchers and Cooperative Park Studies Units that supported graduate student education and research. These operations were moved, with research arms of other DOI agencies, to the short-lived National Biological Survey, and ultimately to the US Geological Survey in 1996. Lastly, a broad reform of the National Park Service was initiated under the *National Park Omnibus Management Act*, 1998. Many national parks had deteriorated substantially through the years and required significant infrastructure modernization. The Omnibus Act reformed the concession management practices within parks. In many parks the concessions included a host of things such as fast food restaurants, souvenir shops, rental of trail horses, and hotels. The Act also instigated visitor fees to National Parks.

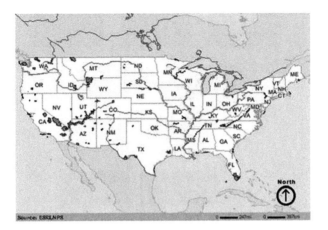

FIGURE 6.14   Major land holdings and National Parks of the National Park Service. *Credit: US National Park Service.*

As of 2008, the NPS included: 84 million acres (34 million ha) of land; 4.5 million acres (1.8 million ha) of oceans, lakes and reservoirs; 85,049 miles (136,869 km) of perennial rivers and streams, 68,561 archeological sites, 43,162 miles (69,460 km) of shoreline, 27,000 historic structures, 2,461 national historic landmarks, 582 national natural landmarks, 400 endangered species, 397 national parks, and 40 national heritage areas.[13]

Organizational Structure of the National Park Service

The Washington Office organizational chart as in 2012 is shown in Figure 6.15. The director has two deputy directors, but the deputy director for operations appears to have far greater responsibility than the director for support services. Under the deputy director of operations are associate directors who oversee programs, assistant directors who primarily deal with business issues, and regional directors who maintain the regional offices across the country. The deputy director for support services, on the other hand, supervises many of the external functions of the Service, including media, outreach to states and tribes and other nations, and education. The policy function of this deputy director involves implementing the service-wide policy set by the director of the Service and his/her supervisors, including those in the Department of the Interior, President and Congress. As with all federal agencies, the National Park Service must develop a strategic plan every five years. These plans evaluate the activities of the previous five years and how they met the goals and objectives set during that time, and establish new goals for the next five years.

The deputy director of operations, along with the comptroller, oversees the business end of the NPS, including human resources, contracts, budget,

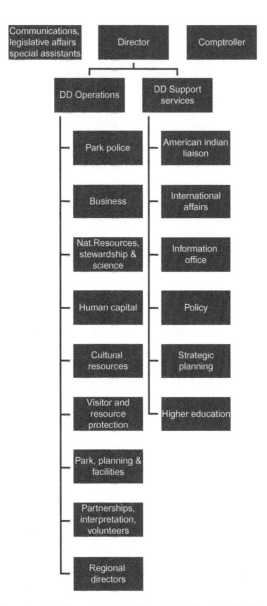

FIGURE 6.15 Organizational Structure of the National Park Service, Washington, DC. *Credit: US National Park Service.*

concessions, and park fees. In addition, the deputy director has oversight of most of the functions of the national parks and historical sites within the service. The Division of Cultural Resources includes the national historical sites, national heritage program, and museum. These sites include everything from Thomas Jefferson's home to the museums on our capitol's mall area. If you

FIGURE 6.16   Regional Service Offices of the National Park Service. *Credit: US National Park Service.*

know of any buildings that have been designated as national heritage, the designation has ultimately been determined by this office. Park planning, facilities and lands covers the operation of the national parks, while the Office of Natural Resources Stewardship and Science oversees the minerals, water, plants and animals on these parks. The offices dealing with visitor and resource protection, park police, and partnerships are involved with people management.

The National Park Service consists of six service regions (Figure 6.16), with Alaska serving as its own region, the Pacific islands falling within the Pacific West Region, and Puerto Rico and the Virgin Islands being included in the Southeast service region. These regions have offices that generally correspond to the Washington DC offices and serve the parks and sites within their geographical boundaries.

National Park Service Budget

The annual budgets requested for the NPS hover around $2.9 billion. In addition to the money requested from the federal budget, the Park Service collects about $190 million from recreational fees, $60 million from park concession fees, and $1.2 million in special use fees each year. The largest portion of the service budget by far is operations (Figure 6.17). This portion of the budget is further divided into six categories corresponding to the major organizational areas of the NPS. Facilities maintenance gets almost one-third of the operations budget; visitor services, stewardship, park support, and park protection have comparable budgets, and external administration gets about one-half that of the other divisions. In addition to operations, major expenditures by the NPS include land acquisition; construction, preservation and recreational costs are comparatively small.

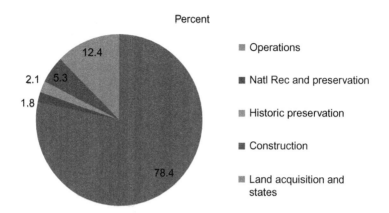

FIGURE 6.17 Distribution of National Park Service Funds by Program. The biggest portion is Operations, which includes running the parks. *Credit: US National Park Service. This figure is reproduced in color in the color plate section.*

BUREAU OF LAND MANAGEMENT

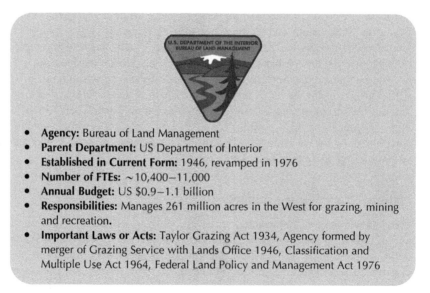

- **Agency:** Bureau of Land Management
- **Parent Department:** US Department of Interior
- **Established in Current Form:** 1946, revamped in 1976
- **Number of FTEs:** ~10,400–11,000
- **Annual Budget:** US $0.9–1.1 billion
- **Responsibilities:** Manages 261 million acres in the West for grazing, mining and recreation.
- **Important Laws or Acts:** Taylor Grazing Act 1934, Agency formed by merger of Grazing Service with Lands Office 1946, Classification and Multiple Use Act 1964, Federal Land Policy and Management Act 1976

The *Bureau of Land Management* (BLM) is the nation's largest federal landholding agency. Altogether, it manages more than 248 million acres (100.4 million ha) of surface land and holds mineral rights on an additional 255.7 million acres (103.5 million ha), almost all of it in the western states. This is equivalent to approximately 12% of the total land mass of the United States.[14] The difference between surface and subsurface rights is due to the laws of individual states. Some states separate mineral rights from surface rights – someone can own the surface property, but someone else can hold the mineral rights under that property. In other states, mineral and surface rights are inseparable. Total subsurface holdings, therefore, is the sum of the surface and subsurface real estate. In some cases the BLM owns a large proportion of a state's total area – for example, the BLM manages two-thirds of Nevada (Table 6.1).[15] East of the Mississippi, the BLM manages around 39 million acres (15.8 million ha).

TABLE 6.1 Number of Acres (in Millions) and Percent of Total State Area Owned by the Bureau of Land Management, by Western State

State	Total area of state	BLM surface	BLM mineral or subsurface	Tribal* lands subsurface	Surface percent of state
Alaska	424	85.6	151.4	1.2	20.2
Arizona	73.0	12.2	23.6	20.7	16.7
California	104.8	15.2	47.5	0.6	14.5
Colorado	66.6	8.4	20.6	0.8	12.6
Idaho	53.5	12	24.5	0.6	22.4
Kansas	52.7	0	0.8	0	0
Montana	94.1	8	29.8	5.5	8.5
Nevada	70.7	47.3	40.9	1.2	66.9
New Mexico	77.8	13.4	22.6	8.4	17.2
North Dakota	45.2	0.059	5.5	0.9	0.1
Oklahoma	44.7	0.02	2.3	1.1	0
Oregon	63.0	16.1	17.8	0.8	25.5
South Dakota	49.5	0.27	3.43	5.0	0.5
Texas	171.9	0.12	4.4	0	0
Utah	54.3	22.9	12.3	2.3	42.1
Washington	45.6	0.44	12.1	0	0.9
Wyoming	62.6	18.4	23.5	1.9	29.4
Totals	1554	260.4	442.6†	51	16.7

*The Bureau of Land Management has agreements with several Native American groups to manage the mineral rights on their reservations or nations.
†Total subsurface land is found by adding surface and subsurface.
Source: Bureau of Land Management[15].

The mission of the Bureau of Land Management is "to sustain the health, diversity, and productivity of America's public lands for the use and enjoyment of present and future generations".[14] The Bureau operates the surface lands under *a multiple use, sustained yield* policy. This means that the bureau allows several different activities – several more than would be allowed by the FWS or NPS – and tries to coordinate them so that they do not interfere with each other. Primary uses of the lands include timber harvesting, mining, livestock grazing, and recreation. Energy development is an important priority, and oil and coal extraction on BLM lands has occurred for decades. Other non-renewable sources of energy on BLM lands include oil shale and tar sand mining. Through leases, extraction fees and other methods, the Bureau receives around $5 billion annually from corporations and ranchers – actually more than its annual federal budget. In addition to the commodities it provides, the BLM has almost 70 million visitor days a year. Because of the astounding amount of land it controls, the BLM provides habitat for over 3000 species of animals, including 140 species of concern.

History of the Agency

The Bureau of Land Management, established in 1946, is a relatively young DOI agency, but it has a historical background as interesting as the Old West. The earliest foundation for the BLM was the *Land Ordinance* of 1785. A year prior to that, Congress determined that the land east of the Mississippi River, north of the Ohio River and west of the Appalachian mountains was to be divided into 10 states, but they did not determine how that was to be done. Under the Land Ordinance, the *Public Land Survey System* was established. This agency determined that all lands under its jurisdiction, and subsequently all lands west of the Appalachians, were to be divided into square townships consisting of 6 miles (9.6 km) a side and oriented on the major compass points. Each square mile was a section, so there were 36 sequentially numbered sections per township. Initially section 16 was set aside for schools, and later section 36 was included in school support. States east of the Appalachians had their own survey systems, and travelers from the Midwest to the East Coast or *vice versa* are often amazed at the regularity the Public Land Survey provides. The Land Ordinance also gave the federal government the right to survey unsettled lands. *The Northwest Ordinance* 1787 established the Northwest Territory, which lay south of the Great Lakes, north and west of the Ohio River, and east of the Mississippi River. It also set a precedent for the United States government to establish territories and encourage settlements. The *General Lands Office* was established in 1812 to handle the land claims filed by settlers.

A few years before the Civil War, there was a growing interest in moving from the east and Midwest towards the unsettled West. Congress encouraged this interest by enacting a series of *Homestead Acts*. Initially (1862), settlers were given 160 acres (65 ha) of land free of charge. This was expanded to

320 acres (130 ha) as prime land was taken, and, by 1909, to 640 acres (259 ha) of grazing land. By the time the Homestead Acts were discontinued, 270 million acres (420,000 square miles or 109 million hectares) had become settled.

However, not all was calm and peaceful out west. Less scrupulous individuals from large cattle operations added to their holdings by staking claims on nearby lands, often to control valuable resources such as water, timber or minerals. Sequestration of water frequently led to range wars as other ranchers demanded equal access to this limited resource. While most of the early settlers were ranchers and favored open grazing land, the Homestead Acts brought many farmers into the area and, with them, fences. Ranchers and farmers did not get along very well – as anyone who has watched an old Western movie can attest.

The *Mining Law* of 1872 aided the Homestead Acts in drawing people into the West. This law gave away mining lands to anyone who filed a claim for the extraction of gold, silver, cinnabar and other valuable minerals. Many of those who came to strike it rich struck out instead, and stayed rather than going home. Today this mining law still allows individuals and corporations to mine royalty-free, to the tune of several million dollars per year, through grandfather clauses.

In the late 19th century, attitudes within the US government began to change. For a while land was still given away, but Congress started to set aside lands as national parks, reserves, forests and refuges, and began to develop a better appreciation for the intrinsic value of these lands and the need to preserve some for posterity. In 1920 Congress enacted the *Mineral Leasing Act*, which began to reverse the land giveaways and retained federal land to lease the rights to extract minerals.

Grazing on BLM lands came under tighter control with the *Taylor Grazing Act* 1934. Prior to this Act much of the western range was deteriorating, as ranchers had no rules and often overgrazed open land. This Act set up grazing districts within the West that were to establish animal units per month (AUM; the number of sheep or cattle that could be grazed per month on a range-specific basis). It also established the Division of Grazing (later *US Grazing Service*) to provide advice and guidance to the grazing districts. In 1946, the Grazing Service was combined with the Lands Office to form the present day *Bureau of Land Management*.

Organization of the Bureau of Land Management

The organization of the BLM is similar to that of the Fish and Wildlife Service and the National Park Service, in that there is a director, deputy directors, sections and regional offices. For that reason we won't print the organization chart, but we'll briefly discuss the more salient points. The BLM has five divisions directly connected to natural resource management: Fire and Aviation; Renewable Resources and Planning; Minerals and Realty

Management; National Landscape Conservation System and Community Partnerships; and Law Enforcement and Security. Additionally, there are five offices that deal with the internal and external business affairs of the Bureau.

Fire and Aviation deals with all things associated with fire, including extinguishing unwanted range fires and planning and implementing prescribed burns. Each year, hundreds to thousands of fires occur within BLM and surrounding lands. Some of these are weather related, while others are accidentally set from poorly tended campfires, tossed cigarettes, or other sources. Prescribed burns help reduce standing fuel levels and thus mitigate the chances of uncontrolled fires. Over a recent five-year period, the Bureau conducted controlled burns on over 2 million acres (0.81 million ha).[16]

Renewable Resources and Planning is the largest of the divisions within BLM. It includes most of the aboveground natural resource activities in the agency, such as fisheries and wildlife, rangeland management, paleontological and cultural activities, forests, environmental quality and wild horses and burros. The wild horses and burros program originated when Congress passed the *Wild Free-Roaming Horses and Burros Act* in 1971. These animals originally came from western settlers and miners. Both the burros and horses sometimes escaped, and on occasion a hard-luck miner would release his burro when he quit his claim. Populations of these animals grew over time and, because the equines represent part of our historical heritage, they were afforded protection through the Act. Periodically the numbers of horses and burros are estimated and the Bureau must decide if they are over their carrying capacity. If so, the populations are thinned, often through private adoption. Currently, some 37,000 horses and burros range over 31.6 million acres (12.8 million ha).[17]

The Minerals and Realty Management deals with subsurface activities and the acquisition or sale of public lands. Mining activities are included in this division. The Office of Law Enforcement is taxed with trying to assure that only legal activities occur on Bureau lands. However, by one estimate, budget restrictions have reduced the law enforcement ranks so that, on average, each ranger covers 1.22 million acres[18] (0.49 million ha). I dare say that no person can cover that much territory effectively.

National Landscape Conservation and Community Partnerships protects areas with the greatest natural beauty; among these are 887 federally recognized areas covering approximately 27 million acres (10.9 million ha) of national monuments, conservation areas, and wilderness areas. The divisions not directly tied to natural resource management include information management, communications, human capital management, business and fiscal resources functions.

The 13 regional offices of the BLM are, in general, the individual states that include Bureau Land. The exceptions to a one-state region include the Eastern States Region, that covers all states with BLM lands east of the

Mississippi River, and the Montana/Dakotas Regional Office, encompassing Montana, and North and South Dakota. By having states as separate regions, management activities can be planned at a local level.

Budget of the Bureau of Land Management

The annual budget for the BLM varies around $0.9–1.1 billion. Direct management of lands and resources accounts for nearly 84% of the total budget (Figure 6.18). Of that total, 27% goes to land resources, 17% to work force management, and 11% to resource protection. The Oregon and California grant program (about $112 million in 2012) reimburses money to 18 counties in Oregon that were affected when the Oregon and California Railroad defaulted in selling about 2.6 million acres (1 million ha) that had been given them by the federal government. The government assumed the obligation of this property exchange and is paying the counties for lost income on timber and minerals.

As mentioned, some of the Bureau's activities, such as mineral rights leasing, grazing rights, and fees and fines, generate money. For example, in 2011, $27,725 million was generated. Of the grazing and mineral fees, 50% is returned to the respective states, 40% is put into a Reclamation Trust, and the Bureau uses the remaining 10% to cover administrative costs.

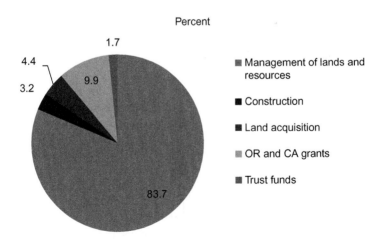

FIGURE 6.18 Distribution of Bureau of Land Management funds by program. *Credit: Bureau of Land Management. This figure is reproduced in color in the color plate section.*

UNITED STATES GEOLOGICAL SURVEY

- **Agency:** US Geological Survey
- **Parent Department:** Department of the Interior
- **Established in Current Form:** 1897, 1995
- **Number of FTEs:** 8500
- **Annual Budget:** US $1.0–1.1 billion
- **Responsibilities:** Provides science services to government agencies and the American people with emphasis on mineralogy, water, cartography and biology. Houses the Cooperative Fish and Wildlife Units; only agency within the US DOI authorized to conduct research.
- **Important Laws or Acts:** Coast and Geodetic Survey enacted 1807, Sundry Civil Appropriations Act 1897.

According to its website, the *United States Geological Survey* (USGS) is "a science organization that provides impartial information on the health of our ecosystems and environment, the natural hazards that threaten us, the natural resources we rely on, the impacts of climate and land-use change, and the core science systems that help us provide timely, relevant, and useable information".[19] Its corresponding mission statement reads, "To sustain the health, diversity, and productivity of America".[19]

As further explanation, the USGS is primarily a provider of scientific information dealing with water and minerals of the United States. It also has an exceptional and renowned cartographic capability, and is the DOI agency for biological and ecological monitoring. From its inception, the USGS has had the primary objectives of surveying, monitoring and cataloging water, mineral and geographic data. After 1993, it added monitoring and surveying of biological information. The agency uses that information to advise other agencies, states and tribes, and often puts the information into databases that can be publically accessed. The USGS should not be confused with an agency that has a primary directive of hypothesis testing or experimentally driven research, although it has some scientists that do that type of work − primarily in the biological discipline of the agency. The USGS is a global authority on volcanology and seismology (the study of volcanoes and movements of the Earth).

The USGS does much of its investigations at the behest and expense of other federal and state agencies. It therefore looks at its cooperators, unusually, as "customers", not collaborators or partners. The USGS receives a substantial portion of its budget through these interagency agreements.

On occasion, this places the USGS in direct competition with other organizations such as universities and non-governmental agencies in looking for funding.

History of the US Geological Survey

The earliest surveys in the United States were performed in the early 1800s. These were primarily agricultural surveys conducted at the county level, although the Army Engineers were commissioned to conduct surveys for roads and canals associated with military and commercial needs. Spurred by the discovery of gold and other precious minerals in the Southeast and West, a drive for a transcontinental railroad, and a need to classify federal land holdings in the 1850s, Congressional interest in surveying the nation grew. Further progress was temporarily halted, however, by the Civil War. In 1867 four major surveys were conducted in the western territories, and these rekindled the desire for more information on the nation's resources. As a result, the US Geological Survey was established in 1879 for the "classification of the public lands, and examination of the geological structure, mineral resources, and products of the national domain".[20] The first duty of the new agency was to classify the public lands, which at that time amounted to 1.2 billion acres (480 million ha) of which only 1% had been surveyed. Because of the extensive area of dry and semi-desert lands west of the Mississippi, considerable effort was devoted to surveying waterways during the 1880s, and this was a precursor to the water division of the USGS. Additional emphasis on water resources occurred during the mid-1890s, to gauge streams and determine the amount of drinking and irrigation water available to the United States. Surveys of underground streams and aquifers started in 1895, and the water division was on its way. Although mapping was a part of the USGS from the start, it too blossomed in the 1890s when Congress called for a topographical map of the United States.

The emphasis on external funding for the USGS seems to have begun in the early 1900s. For 20 years, under Director George Otis Smith, the direction of the USGS took a distinct turn towards the practical and pragmatic. Federal funding was more or less static during this time, but revenues from water resource programs and topographic mapping increased appreciably. Outside funding increasingly became important – particularly during the depression, when many federal budgets were slashed – but the products of the USGS were still in demand. As one example, the Tennessee Valley Authority, a quasi-government agency, contracted with the USGS to develop topographical maps for the entire valley and for an expanded stream-gauging program in 1933. Also, the Public Works Administration contracted with the USGS to conduct mapping and water projects that totaled $3.7 million (over $6.2 billion by today's value) – more than the federal allocation to the

USGS that year. In 1933, funding from other agencies was four times greater than the Congressional appropriations. At its 75th anniversary in 1954, the USGS had a budget that included federally appropriated funding of $27 million and an external funding base of $48 million. Clearly, external funding was lucrative. The USGS has had a long and successful track record of soliciting funds.

World War II was an important time for the USGS. Its cartographic branch was called into service to produce military-quality maps of Alaska and foreign lands. The minerals branch was asked to locate extractable sources of needed metals and minerals, including petroleum. Water resources staff provided information on water sources and supply both nationally and internationally. New techniques for finding and extracting these resources were developed, and the USGS flourished. Growth of the agency continued after the war as the need for metals to replenish exhausted stores, and for mapping and water resources for a population continuing to move into the arid West, increased. The nuclear age produced a need for rare earths and other metalloids that had not been in much use prior to the war, and the agency became involved in locating deposits.

The late 1950s and 1960s welcomed geological research as the agency was called to determine the effects of underground nuclear testing and begin its volcanology program in the Hawaiian Islands. Earth-based USGS assisted NASA with its early Man in Space programs by finding lunar-like areas on this planet that could serve as training sites. The great Alaskan earthquake of 1964 initiated a longstanding program in earthquake prediction and monitoring; this program was later expanded to include additional natural disasters. Today the USGS earthquake network is global, and the agency is the world's authority for determining the epicenter and strength of earthquakes.

Over the course of its history, various segments of the USGS broke away from their parent and formed new agencies that still exist. These changes attest to differences in policy within the federal government. In 1891 the forest reserves were placed under the Department of the Interior, and subsequently the USGS. However, in 1905 these tracts were moved into the US Forest Service. In 1902, the USGS was given the authority of water management in the West. Five years later this function was moved into the new Bureau of Reclamation. That same year the Technologic Branch was formed to test fuels and structural materials within the USGS, and three years later the Bureau of Mines was formed from this group. Congress had given the USGS the authority to classify grazing and agricultural lands in 1909, but when the Taylor Grazing Act was passed in 1934 this function was spun off into the Grazing Service of the US Department of Agriculture. In 1982, under the Reagan administration, the longstanding Conservation Branch of the USGS was moved into the newly formed Minerals Management Service, which has subsequently become the Bureau of Ocean Energy Management (BOEM) and the Bureau of Safety and Environmental Enforcement (BSEE).

In the early 1990s the Secretary of the Interior, Bruce Babbitt, became concerned because he perceived that ecological research within the federal government might be tainted by agency perspectives. For example, he felt that the scientific opinions on old-growth forests in the Pacific Northwest reflected the attitudes of the agencies involved, including the Forest Service, Fish and Wildlife Service, and Bureau of Land Management. Because Babbitt rightfully believed that research should be as objective and as free from agency policy as possible, he created a new agency, the National Biological Survey (also known as the Service, or NBS), which was to house all research biologists and their facilities within the Department of the Interior. Unfortunately, full authority to create such an agency needed legislative support, which was not obtained beforehand. Congress took a dim view of this and declared that the NBS would have to be dismantled. As a result, all research within the DOI was transferred to the USGS in 1996. Many of the scientists who were transferred into the USGS at this time had a long history of experimental research, rather than monitoring and surveying. In addition, most had relied on funding from within their respective agencies to conduct their studies rather than hustling to find someone else to sponsor projects. Their ability to find funds was further hampered by other factors — for example, federal agencies generally are not allowed to seek grants from the National Science Foundation or National Health Service, and states are often very reluctant to give their money to federal agencies. As a result, several researchers (including the author) left the USGS for other employment, and the nature of biological research within the DOI has been drastically altered.

Organizational Structure of the US Geological Survey

The administrative lines within the USGS are remarkably flat (Figure 6.19). Essentially, all major offices report directly to the director and deputy director. For most of its history, the USGS's organization was divided into discipline branches (or bureaus, offices, departments or disciplines depending on time period). Until 1996, when Biology was added, these primarily consisted of Cartography, Minerals, and Water. Under this structure, biologists, hydrologists, mineralogists and map makers were generally in different offices and seldom conferred with each other. Recently, the USGS reconfigured its structure to a "mission area" approach with the concept that in some areas more can be accomplished by scientists from multiple disciplines working together than by specific disciplines working separately. Findings from the various mission areas are evaluated by a Science and Decisions Center, which cross-cuts the various disciplines to provide integrated assessments of the environment.

The Ecosystems mission area is primarily biological. It includes groups studying invasive species, fisheries, ecosystem restoration and adaptive

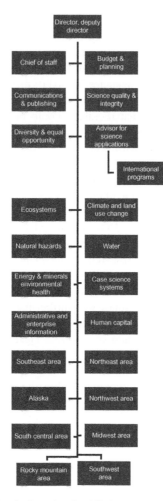

FIGURE 6.19 Organizational chart for the US Geological Survey. *Information from US Geological Survey.*

management, wildlife, genetics, microbiology, the population status and trends of wildlife and fish, natural resource policy, documentation of seasonal patterns of living organisms, and the interaction between wildlife and renewable energy conservation.

This Ecosystems area also includes the *Cooperative Research Units*. Each of these 40 units housed in 38 states (Figure 6.20) is formed by cooperation among the USGS, a university, at least one state conservation agency, and the Wildlife Management Institute. The USGS provides the unit leadership, the university offers lab and office space and faculty status to the leaders, one or more state agencies provide financial assistance for research projects and a

Alaska
Hawaii
Headquarters

FIGURE 6.20 Distribution of Cooperative Fisheries and Wildlife Units of the US Geological Survey across the United States. These units have been instrumental in the education of wildlife and fisheries graduate students for many decades. *Credit: US Geological Survey.*

secretary, and the Wildlife Management Institute provides some funding. Most of these units were transferred from the US Fish and Wildlife Service, while others came from the National Park Service. The purpose of the cooperative units is to provide graduate-level training and education in conservation, wildlife and fisheries biology, and management. Degrees are conferred by a host of academic departments, but the research occurs through the unit. The program has been very successful in producing highly competent scientists.

The Climate and Land Use Change mission combines scientists from different disciplines to study climate change and ways to mitigate its effects. Cartographers model what may happen with sea-level rise; they are helped by geographers using geographic information systems (GIS). Biologists look at possible changes in animal distributions, and other scientists use remote sensing, including satellite monitoring of the Earth's surface, to study patterns and trends.

The directorate for Natural Hazards, consisting mostly of geologists, is a leading global player in the study and prediction of earthquakes, tsunamis, volcanoes, coastal geology, geomagnetism and landslides. It operates a global seismographic network for the study of shifts in the Earth's crust that could lead to natural disasters. The Water mission area is staffed primarily by hydrologists who monitor the nation's water systems, assess and maintain databases on water quality, continue the early USGS work of gauging streams, and work internationally to help locate high-quality water for irrigation and human consumption.

The *National Water Quality Assessment Program* (NAWQA) is the largest continual monitoring program of streams, rivers and ground water quality in the nation. It consists of both monitoring or data collection

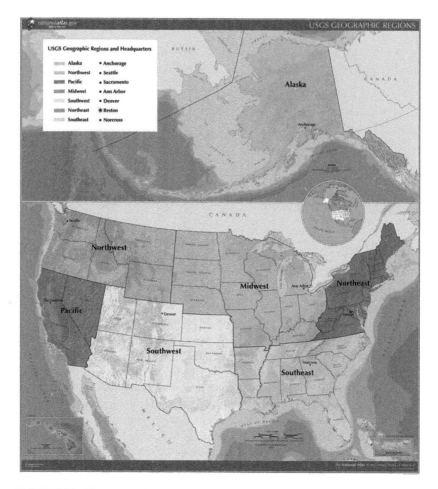

FIGURE 6.21 The seven regions of the US Geological Survey. *Credit: US Geological Survey.*

activities and a national reporting database that holds thousands of records on tissue, water and sediment-borne contaminants.

The Energy, Minerals, and Environmental Health mission area is another section that relies on multiple disciplines. It consists of groups that study the sources and occurrence of contaminants, and conduct research on the effects of these contaminants in living organisms. There are also groups that investigate renewable and non-renewable energy and mineral sources and the impact that utilization of these resources might have on humans and the environment. Core Science Systems is where you'll find most of the cartographers. Under a recent reorganization the USGS is divided into seven regions corresponding to major sections of the country (Figure 6.21).

Percent

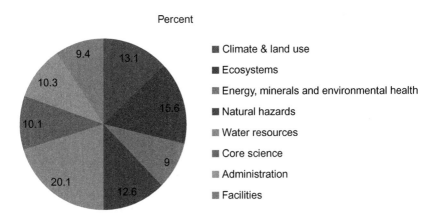

- Climate & land use
- Ecosystems
- Energy, minerals and environmental health
- Natural hazards
- Water resources
- Core science
- Administration
- Facilities

FIGURE 6.22   Annual budget by program for the US Geological Survey. *Information from US Geological Survey. This figure is reproduced in color in the color plate section.*

The US Geological Survey Budget

Recall that the USGS receives substantial payment from states, other federal agencies and other organizations for its services. In addition, it receives approximately $1.1 billion in federal appropriations each year. Other than Water Resources, which receives about 20% of the annual budget, the other mission areas, administration and facilities receive nearly equal proportions of the annual budget (Figure 6.22). In a recent year the USGS took in $447.1 million, of which 49.5% came from other federal agencies, 46.5% from local and state government, and 4% from other sources.

OTHER DEPARTMENT OF THE INTERIOR AGENCIES

In addition to the agencies discussed above, the DOI has five agencies that are less focused on natural resources, more focused on non-resource issues, or more specialized. They are mentioned here to complete our discussion of the Department.

Bureau of Indian Affairs

The Bureau of Indian Affairs (BIA) is one of the largest agencies within DOI, and was established in 1824 to "deal" with Native Americans. There are many examples of abuse and injustice afflicted upon the Native Americans, and it is well known that in the early history of the United States Native Americans were considered a direct impediment to the development of the West. As a result, they were constantly moved from one area to another one further west, and further into desert or semi-desert habitat. Many Native Americans rebelled against these forced movements and abuse, and rose up during the "Indian wars".

Gradually, tensions between the US government and Indian tribes eased. Through actions such as *The Indian Citizenship Act* of 1924 and the *New Deal and the Indian Reorganization Act* of 1934, Congress worked with Native Americans to establish modern tribal governments. Through the *Indian Self-Determination and Education Assistance Act* of 1975 and the *Tribal Self-Governance Act* of 1994, American Indians and Alaskan natives have been granted considerable self-regulatory status. Native American nations such as the Arapaho and Apache nations are independent entities with their own sets of laws and governance; consequently, the BIA has evolved from an agency with the primary responsibility of keeping Native Americans "in check" to one that serves as a conduit for government-to-government cooperation. Its current mission statement is "to enhance the quality of life, to promote economic opportunity, and to carry out the responsibility to protect and improve the trust assets of American Indians, Indian tribes and Alaska Natives".[21]

With regard to natural resources, the BIA provides assistance and guidance on the approximately 55 million surface acres (22 million ha) and 57 million subsurface acres (23 million ha) held in reservations and nations. While maintaining Native American control, the BIA advises on agricultural, rangeland, water rights, licensing of hydroelectric power, and outdoor recreational opportunities. The Bureau consists of 12 regions covering the entire nation, but with primary emphasis in the West. It operates with an annual budget of about $2.5 billion and a staff of 8500, 87% of which is Native American.

Bureau of Ocean Energy Management, Regulation and Enforcement, and Bureau of Safety and Environmental Enforcement

A year after the British Petroleum Deep Water Horizon oil spill of 2010, the Minerals Management Service of the Department of the Interior was divided into the Bureau of Ocean Energy Management, Regulation and Enforcement (BOEM) and the Bureau of Safety and Environmental Enforcement (BSEE). BOEM has the responsibility of managing the development of the nation's offshore resources in an environmental and economic manner. It serves as a watchdog for offshore leasing, resource evaluation, administration of oil and gas exploration, renewable energy development, and environmental studies. It operates in three regions: the Gulf of Mexico, Pacific Ocean, and Alaska. In 2012, its annual budget was $358 million and it had a staff of 1417 FTEs.

BSEE "works to promote safety, protect the environment, and conserve resources offshore through vigorous regulatory oversight and enforcement".[22]

It oversees offshore drilling and mineral exploration, making sure that companies adhere to safety regulations. It also coordinates remediation efforts in case an oil spill occurs, and trains responders on how to handle such spills. Its regions are the same as BOEM, and its annual budget in 2012 was $76 million with a staff of 703.

Bureau of Reclamation

Water is one of the most precious natural resources in the West. Water is used for a growing human population, irrigation of desert sands into a rich agricultural region, and providing for thousands of head of livestock. Water shortages in the late 1800s and early 1900s contributed to range wars as ranchers fought for limited water rights. Around the beginning of the 1900s, only a few hundred thousand people dwelt in the arid southwest of Arizona, California, Nevada and New Mexico; today, about 60 million people live there.[23] Two rivers, the Rio Grande and the Colorado, must provide the bulk of the water for this region, and any serious drought could be disastrous. These are the realities that confront the Bureau of Reclamation (BOR).

The roots for BOR began in 1902, under the US Geological Survey, to study reclamation projects in each western state that had federal lands. At that time "reclamation" meant "reclaiming" desert lands for human use, not reclamation of mined lands as the term is frequently applied today. In 1907, the Bureau was made a separate agency under the Department of the Interior. Over the years, BOR has been responsible for hundreds of water projects including irrigation systems and dams. For example, the Hoover Dam was authorized in 1928 on the Colorado River and the Grand Coulee Dam on the Colombia River in the Pacific Northwest (Figure 6.23), and placed under BOR. During the ecological movement of the 1960s and 1970s, BOR received

FIGURE 6.23 The Grand Coulee Dam on the Columbia River in Washington, one of the many dams managed by the Bureau of Reclamation. *Credit: US Bureau of Reclamation.*

considerable criticism for its stance on damming rivers. In the 1990s, BOR underwent a change from constructing new water projects to managing the facilities that had already been built. Today the mission of the BOR is to "manage, develop, and protect water and related resources in an environmentally and economically sound manner in the interest of the American public".[24] It manages about 180 facilities in 17 western states. Reclamation is the second largest producer of hydroelectric power in the western United States. Its power plants provide more than 40 billion kilowatt hours per year, generating nearly $1 billion in revenues and producing enough electricity to serve 3.5 million homes. Irrigation projects serve 10 million acres (4.8 million ha). The agency has five regions, all west of the Mississippi River, and an annual budget of around $1.1 billion.

Office of Surface Mining Reclamation and Enforcement

Before the 1970s, coal mining companies could come into an area, dig large pits to extract coal, and leave without any remediation. Most often the companies would make huge spoil banks with the original topsoil on the bottom and deeper soil layers progressively on top, following the sequence of digging. Oxidation of the lower soil horizons would cause acidification, which could result in dead zones around the mined areas. To combat this, Congress enacted the Office of Surface Mining Reclamation and Enforcement, often just called the Office of Surface Mining (OSM).

OSM "is responsible for establishing a nationwide program to protect society and the environment from the adverse effects of surface coal mining operations, under which OSM is charged with balancing the nation's need for continued domestic coal production with protection of the environment".[25] The Bureau was created in 1977 under the *Surface Mining Control and Reclamation Act*. OSM works with State and Indian Tribes to assure that people and the environmental are protected during coal mining activities, and that the mining companies reclaim mined lands according to a set of federal regulations. The agency is also responsible for restoring lands mined before 1977, so-called pre-law mines, most of which were abandoned. Initially, OSM directly enforced mining laws and arranged clean-up of abandoned mine lands. Today, the agency mostly works with states and tribes in helping enforce their own programs. The agency has three regions: Appalachian, Mid-continent and Western. Its annual budget is around $685,000, with a fulltime workforce of 528.

Study Questions

6.1. What agencies or bureaus are found within the Department of the Interior? Briefly describe the major functions of each.

6.2. Name and describe some of the major laws or Acts that fall under the jurisdiction of the US Fish and Wildlife Service.

6.3. Identify at least three agencies or bureaus that were precursors to the US Fish and Wildlife Service.

6.4. Outline the subdivisions of the US Fish and Wildlife Service that deal with natural resources. What are their specific functions?

6.5. Describe some of the entities under the National Park Service's care. What do they manage besides National Parks?

6.6. What are the major categories of expenditures for the National Park Service?

6.7. If you compare the gross budget of the Bureau of Land Management to the number of acres under its management, it comes out to about $4.21 per acre. Do you think that this is sufficient for proper management? Why do you suppose the budget is so low?

6.8. Have you ever given any thought to the nature of federal research? Should the federal government fund its own studies, or is it OK to seek money from other agencies?

REFERENCES

1. US Department of the Interior. 2013. Who we are. <www.doi.gov/whoweare/interior.cfm>.
2. US Fish and Wildlife Service. Who we are. <www.fws.gov/who>; 2013.
3. US Fish and Wildlife Service. Organizational Chart. <www.fws.gov/offices/orgcht.htm>; 2013.
4. Defenders of Wildlife. <http://www.defenders.org/habitat/refuges/map/map.html>.
5. US Fish and Wildlife Service. National Wildlife Refuges. <www.fws.gov/refuges/about/welcome.html>; 2013.
6. US Fish and Wildlife Service. Migratory Birds. <www.fws.gov/migratorybirds/RegulationsPolicies/mbta/mbtandx.html>; 2013.
7. US Fish and Wildlife Service. Fisheries Program. <www.fws.gov/fisheries/whatwedo/fisheries_farc_program.html>; 2013.
8. US Fish and Wildlife Service. About us. <www.fws.gov/le/AboutLE/about_le.htm>; 2013.
9. US National Park Service. <www.nps.gov>; 2013.
10. US National Park Service. Park History Program. <www.cr.nps.gov/history>; 2013.
11. Act to Establish the National Park Service, 1916.
12. US National Wild and Scenic Rivers System. <http://www.rivers.gov/rivers/>
13. US National Park Service. About us (<www.nps.gov/aboutus/index.htm.>; 2013.
14. Bureau of Land Management. <www.blm.gov>.
15. Bureau of Land Management. About us. www.blm.gov/pgdata/etc/medialib/blm/wo/Communications_Directorate/general_publications/rewards/2005.Par.76390.File.dat/PR05NATLtxt.pdf.
16. US Bureau of Land Management. Fire Fuels Management. <www.blm.gov/nifc/st/en/prog/fire/fuelsmgmt.html>; 2013.
17. US Bureau of Land Management. Wild Horses and burros herd management. <www.blm.gov/wo/st/en/prog/whbprogram/herd_management.html>; 2013.
18. US Bureau of Land Management. Law Enforcement. <http://www.blm.gov/wo/st/en/prog/more/law_enforcement.html>.
19. US Geological Survey. <www.usgs.gov>.
20. Organic Act of the US Geological Survey, 43 U.S.C. 31 *et seq.*
21. US Bureau of Indian Affairs. Who we are. <www.bia.gov/WhoWeAre/BIA/index.htm>; 2013.

22. Bureau of Safety and Environmental Enforcement. <www.bsee.gov/About-BSEE/index.aspx>.

23. Parker, K. Population, Immigration, and the Drying of the American Southwest. <http://www.cis.org/southwest-water-population-growth)>; 2010.

24. US Bureau of Reclamation. <www.usbr.gov/history/borhist.html>.

25. Office of Surface Mining. osmre.gov/aboutus/Aboutus.shtm.

Chapter 7

US Department of Agriculture

Terms to Know

- Morrill Act 1862
- Hatch Act 1877
- Smith-Lever Act 1914
- Agricultural Act 1949
- Forest Reserve Act 1891
- Organic Administration Act 1897
- Weeks Act 1911
- National Forest Multiple Use Sustained Yield Act 1960
- Forest and Rangeland Renewable Resources Planning Act 1974
- Healthy Forests Restoration Act 2003
- Soil Conservation Districts
- Soil Bank Program
- Environmental Quality Incentives Program
- Conservation Reserve Program
- Wetland Reserves Program
- Wildlife Habitat Incentives Program
- Mid-Contract Management
- Swampbuster
- Sodbuster

AN OVERVIEW OF THE US DEPARTMENT OF AGRICULTURE

- **Department:** US Department of Agriculture
- **Established in Current Form:** 1862
- **Number of FTEs:** Approximately 120,000

D.W. Sparling: Natural Resources Administration. DOI: http://dx.doi.org/10.1016/B978-0-12-404647-4.00007-6

- **Annual Budget:** $145–155 billion
- **Responsibilities:** Extensive duties include agricultural and natural resource conservation, food programs, nutritional programs, agricultural and natural resource research, and many others. Primary agencies dealing with natural resource management include US Forest Service and Natural Resource Conservation Service.
- **Important Laws or Acts:** Morrill Act 1862, Hatch Act 1887, Smith-Lever Act 1914, Agricultural Act 1949, various Farm Bills since 1996, and many, many more.

The Department of Agriculture was initially established by President Abraham Lincoln as an independent agency in 1862, with subcabinet status, through the *Morrill Act*. This Act also provided federal lands to any state which intended to establish a college or university that included the instruction of agriculture; this was the start of the "land grant" colleges that today are found in every state. In 1889, the Department was elevated to true departmental status as the *US Department of Agriculture* (USDA). Its mission is to "provide leadership on food, agriculture, natural resources, and related issues based on sound public policy, the best available science, and efficient management".[1] The USDA has many functions associated with farm programs, nutrition, welfare and natural resources, and has nearly twice the number of employees and 7.5 times the annual budget of the Department of the Interior. However, of the total budget of the USDA, only 19% is discretionary funds − i.e., those that the agency can decide how to spend. The other 81% are fixed expenses predetermined to go to various food and nutrition services, farm price supports, cost matches, and other purposes mandated by Congress.

A few notable Congressional Actions of the hundreds that have shaped the USDA since its inception include the *Hatch Act* of 1877, which, along with the Morrill Act, provided funding for the establishment of land grant colleges. Simply put, the Morrill Act dedicated funds and the Hatch Act established a plan for their distribution across the states. In addition to agriculture, these colleges were mandated to teach natural resource management. Many of these land grant colleges are numbered among the largest and most prestigious public universities in the country, including Cornell, Purdue, Iowa State, Michigan State, and many others. In fact, most state universities that are denoted by name of "State University" are land grant colleges, but some, like the University of Illinois, go by other names. Today, including the 31 Native American colleges and several states with multiple land grant colleges, there is a total of 106 such institutions. Many of these universities not only have good agricultural and natural resource programs; their sports teams are often pretty good as well.

The *Smith-Lever Act* 1914 established the agricultural extension system for educating the American public in agriculture and home economics outside of a formal university setting. Land and water conservation was always included in this instruction. From an economic perspective, the *Agricultural Act* 1949 was instrumental in establishing crop support and farm subsidiary programs that allowed the American farmer to be more competitive in a world market and helped preserve the family farm system.

The USDA has seven major program areas overseen by undersecretaries (Figure 7.1). The Natural Resources and Environment program includes the two principal agencies dedicated to natural resource management: the US Forest Service (FS) and the Natural Resources Conservation Service (NRCS). In The Farm and Foreign Agricultural Services, the Farm Service Agency provides the financial administration for the very important Conservation Reserve Program (CRP) which will be discussed shortly. Although only a relatively small part of the USDA (Figure 7.2), the NRCS and FS are very important in the management of natural resources within the United States and will receive nearly exclusive focus in this chapter.

US FOREST SERVICE

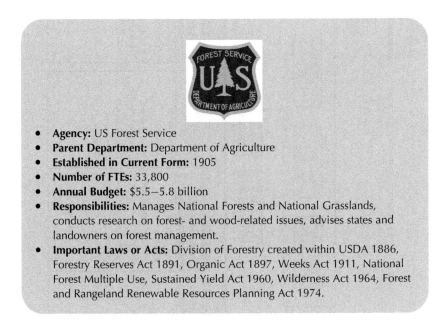

- **Agency:** US Forest Service
- **Parent Department:** Department of Agriculture
- **Established in Current Form:** 1905
- **Number of FTEs:** 33,800
- **Annual Budget:** $5.5–5.8 billion
- **Responsibilities:** Manages National Forests and National Grasslands, conducts research on forest- and wood-related issues, advises states and landowners on forest management.
- **Important Laws or Acts:** Division of Forestry created within USDA 1886, Forestry Reserves Act 1891, Organic Act 1897, Weeks Act 1911, National Forest Multiple Use, Sustained Yield Act 1960, Wilderness Act 1964, Forest and Rangeland Renewable Resources Planning Act 1974.

The US Forest Service (FS) manages about 192 million acres (78 million ha) of forests and grasslands in 44 states, Puerto Rico and the Virgin

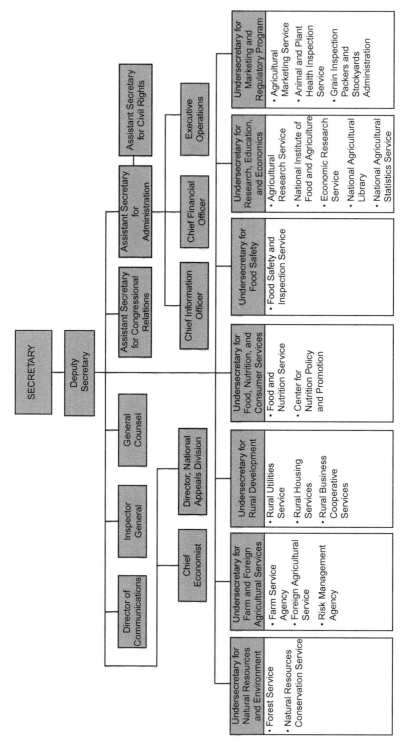

FIGURE 7.1 Organizational chart for the US Department of Agriculture. *Credit: US Department of Agriculture.*

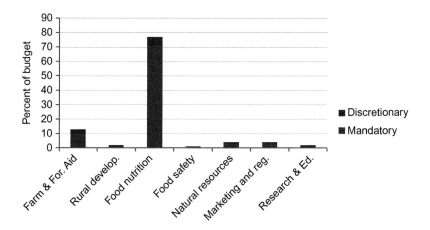

FIGURE 7.2   US Department of Agriculture budget by percent of total budget; both mandatory and discretionary budgets are shown. *Credit: US Department of Agriculture.*

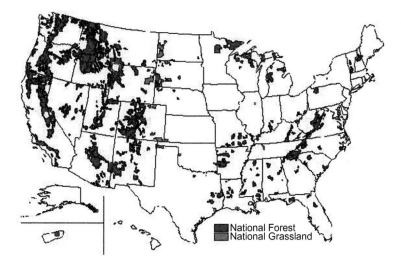

FIGURE 7.3   Distribution of National Forests and Grasslands of the US Forest Service. *Credit: US Forest Service.*

Islands, although its principal holdings are west of the Mississippi River (Figure 7.3). Within that acreage are 155 National Forests, 20 National Grasslands, and most of the designated wilderness areas in the country. Through management of these areas, the FS provides habitat for about 50% of the big-game and cold-water fishes in the United States. It manages, via a multiple use, sustained yield policy, by balancing many different types of activities in the National Forests and Grasslands. The mission of the FS is "to sustain the health, diversity, and productivity of the Nation's forests and grasslands to meet the needs of present and future generations".[2]

History of the US Forest Service

In 1799, the Revolutionary War with Great Britain, which lasted for about seven years, had been over for 16 years, but it was obvious that Great Britain was not content and that the drums of war could sound again. This time, because there was little support for Great Britain within the new United States, the antagonist would have to cross the ocean in ships to attack (or come down from Canada in the north). So Congress began to build up stockpiles of materials. Chief among these stockpiles would be sufficient wood and tall timbers to build the naval ships necessary to repel an attack. The US Congress therefore passed a bill to release $200,000 for the purchase of the US Navy Forest Reserves (today's equivalent = $2.5 million). These reserves established an early precedent for the US Forest Service, although that agency wouldn't be established for another 100 years. Note that this Congressional action was really directed to potential national defense, not conservation *per se*, because Americans still thought that natural resources did not need to be conserved under the mistaken Myth of Superabundance.

Later, as resources began to dwindle, however, some far-seeing individuals began to raise the alarm. In 1864, *George Perkins Marsh* wrote the book *Man and Nature: Or Physical Geography as Modified by Human Action*, in which he likened the current state of resource destruction and overuse to earlier destruction of the Earth and previous civilizations. This influential book drew on the past to illustrate how human actions had harmed the Earth, leading to the demise of earlier civilizations. Marsh was joined by *John Wesley Powell*, and others, who likewise began to sound the alarm. These voices eventually led to the formation of Yellowstone National Park, as explained in Chapter 6.

By the early 1870s, *Dr Franklin B. Hough* noticed that timber production in the East had fallen off in some areas, most of the old-growth forests east of the Ohio River Valley had been clear cut, and timber cutting was progressing westward. This led Hough to believe that timber supplies in some areas of the United States could be exhausted. Hough presented his idea to the American Association for the Advancement of Science, and he was soon commissioned to develop a more complete review of forests and forestry practices in the United States. Eventually, a temporary agency with no enforcement capabilities, the *Department of Agriculture's Division of Forestry*, headed by Hough, was established to further study forestry in America and abroad. *Bernhard Fernow*, who had undergone formal training in forestry in Germany because there was as yet no training available in America, took over the temporary Division in 1886, and one of his principal agenda items was to make the division permanent.

Forests continued to decline because of misuse and lack of regulation, so *the Forest Reserve Act* was established in 1891. This Act allowed the President to set aside forest reserves on public lands; recreation was allowed but logging was not. With this Act, the first National Forest, the Shoshone, was initiated in Wyoming. These lands fell within the jurisdiction of the Department of the Interior. By 1893 President Harrison had established 15 reserves totaling 13 million acres (5.3 million ha) in the western United States. President Cleveland added a further 5 million acres (2 million ha) that same year. However, this was during the historical Era of Preservation and Congress did not make any provision to manage these lands, so the forests were preserved intact.

Further pressure on Congress by leading environmentalists of the time led to the creation of the *Organic Administration Act* of 1897 that defined the function of these forest reserves: "No national forest shall be established, except to improve and protect the forest within the boundaries, or for the purpose of securing favorable conditions of water flows, and to furnish a continuous supply of timber for the use and necessities of citizens of the United States …".[3] If mining was already established on these forest reserves, it could continue. At last, the United States had millions of acres of forest tracts that could be used and managed for timber, recreation, and other activities. Arguably, the economic values of most forests are greatest in the mid-successional stages, and active management helps maintain these values. Shortly thereafter, the Forest Reserves were transferred to the US Department of Agriculture as the Bureau of Forestry; this became the *US* Forest Service in 1907, and the Forest Reserves were renamed *National Forests.*

The first director or *Chief Forester* of the US Forest Service was *Gifford Pinchot* (Figure 7.4), sometimes referred to as the "Father of Modern Forestry". Pinchot came from a family that obtained its wealth from lumbering and land speculation. He received his bachelor's degree from Yale in 1889 and spent some time at the French National School of Forestry, for there still wasn't a forestry school in the United States. The European style of forest management that Pinchot brought with him to the US Forest Service was based on a multiple use, sustained yield model. We encountered this philosophy when we discussed the Bureau of Land Management in Chapter 6. In forests, multiple use means including recreation, grazing, mining, and a host of other activities; sustained yield means conserving the forests to maintain a natural replacement of timber. By choosing this route Pinchot intentionally placed himself between the commercial logging companies, who only saw the economic value of forests, and environmentalists such as John Muir, who wanted forests to be preserved, not conserved. Many people have compared the accomplishments of Gifford Pinchot in forestry with those of Aldo Leopold in wildlife.

Up until this time, only lands west of the Mississippi River could be converted into National Forests; the Service did not have jurisdiction over any forests in the East. While the West had the largest tracts of forested land,

FIGURE 7.4   Gifford Pinchot, known as the Father of Modern Forestry. *Credit: US Forest Service.*

there were still huge, largely unprotected forests east of the Mississippi (Figure 7.3). In 1911 Congress passed the *Weeks Act* and gave the Forest Service the authority to acquire lands in the East. The Service leveraged its budget by setting up a matching grant system with states so that more land could be acquired than the FS alone could afford. Many of the early purchased blocks of forest were patchworks, as some private landowners sold their properties while others did not. Over the years, many of these missing pieces have been filled in. The first purchase of an eastern National Forest consisted of 18,500 acres (7500 ha) in North Carolina in 1916, which became the Pisgah National Forest. Today, through additional purchases, the Pisgah is over 500,000 acres (202,500 ha). By 1940, Eastern National Forests contained about 24 million acres (9.7 million ha).

Although Gifford Pinchot introduced multiple use to US forests in the early 1900s, this management method was officially engrained into the operation of National Forests and National Grasslands through the *National Forest Multiple Use Sustained Yield Act* in 1960. The chief multiple uses defined by this Act included outdoor recreation, grazing, timber, watershed, wildlife and fish. This Act was instituted in part because Congress deemed that too much emphasis was being placed on timber resources and not enough on other values of National Forests. To aid with the management of National Forests, Congress enacted the *Forest and Rangeland Renewable Resources Planning Act* in 1974, which mandated that supervisors of National Forests develop 5- and 10-year plans for their management. At the time mandated conservation plans were new, but subsequently almost all federal and state natural resource agencies instituted similar laws requiring that they develop and adhere to comprehensive plans. Such plans are open to public comment because of NEPA.

Organization of the US Forest Service

The US Forest Service is headed by the Chief Forester who is abetted by a staff for law enforcement, climate change, external affairs, civil rights, finances, and international programs (Figure 7.5). Also directly answerable to the Chief are deputy chiefs in charge of business, the natural resource programs: national forests; research and development; and state and private forestry. The Chief also has in-line responsibilities for the Forest Service Regions, Experimental Stations, labs and institutes, and state and private forests in the northeastern United States.

As with other agencies, the Forest Service must have certain business-oriented offices that take care of the daily and annual processes of human resources, budget, procurement, planning and safety, and the interactions of the agency with other nations and other agencies. Often the personnel in these offices do not have much professional contact with the resources themselves; they may be lawyers, accountants, office support staff, and other administrators. Also, the FS contains nine regional offices that carry out the functions of the agency on a more local level than the national office (Figure 7.6). A common question is, if the FS has nine regions, then how come we have a Region 10 in Alaska? The answer is that in 1965 the current Eastern Region was formed by combination of the Eastern and Northern Regions.[4] They just never got around to renumbering the remaining regions.

The Deputy Chief for the National Forest System oversees the major involvement of the FS with natural resources. The Deputy Chief's job is to provide "big picture" management of all the national forests, grasslands, wild and scenic rivers, and rangelands, as well as the fish, wildlife, mineral extraction, timber, watersheds, historical sites and ecosystem management on these lands. Forest Service Engineering does what its name implies; it provides direction and assistance to the Forest Service and other federal agencies for building roads, bridges, dams, facilities and the like. The Lands office is the real estate office for the FS, dealing with the acquisition and selling of lands. Wilderness and Wild and Scenic Rivers manages over 36 million acres (14.6 million ha) of wilderness; 1.2 million miles (around 2 million km) of wild and scenic rivers; 898,000 acres (363,690 ha) of recreation areas in three major units; and assorted historic sites, monuments and other protected areas. A substantial portion of the total National Forest acreage, 96 million acres (39 million ha), is in rangeland, which is managed from the Office of Rangelands. This includes the National Grasslands plus all of the rangeland within National Forests. Like the BLM, the Forest Service has an obligation to manage the quality of rangelands, allocate animal use units, collect grazing fees, and manage the wild horses and burros on its lands to ensure that their numbers do not exceed their carrying capacity.

The rest of the National Forest acreage is managed by the Forest Management Office. This office oversees timber sales, and makes sure that

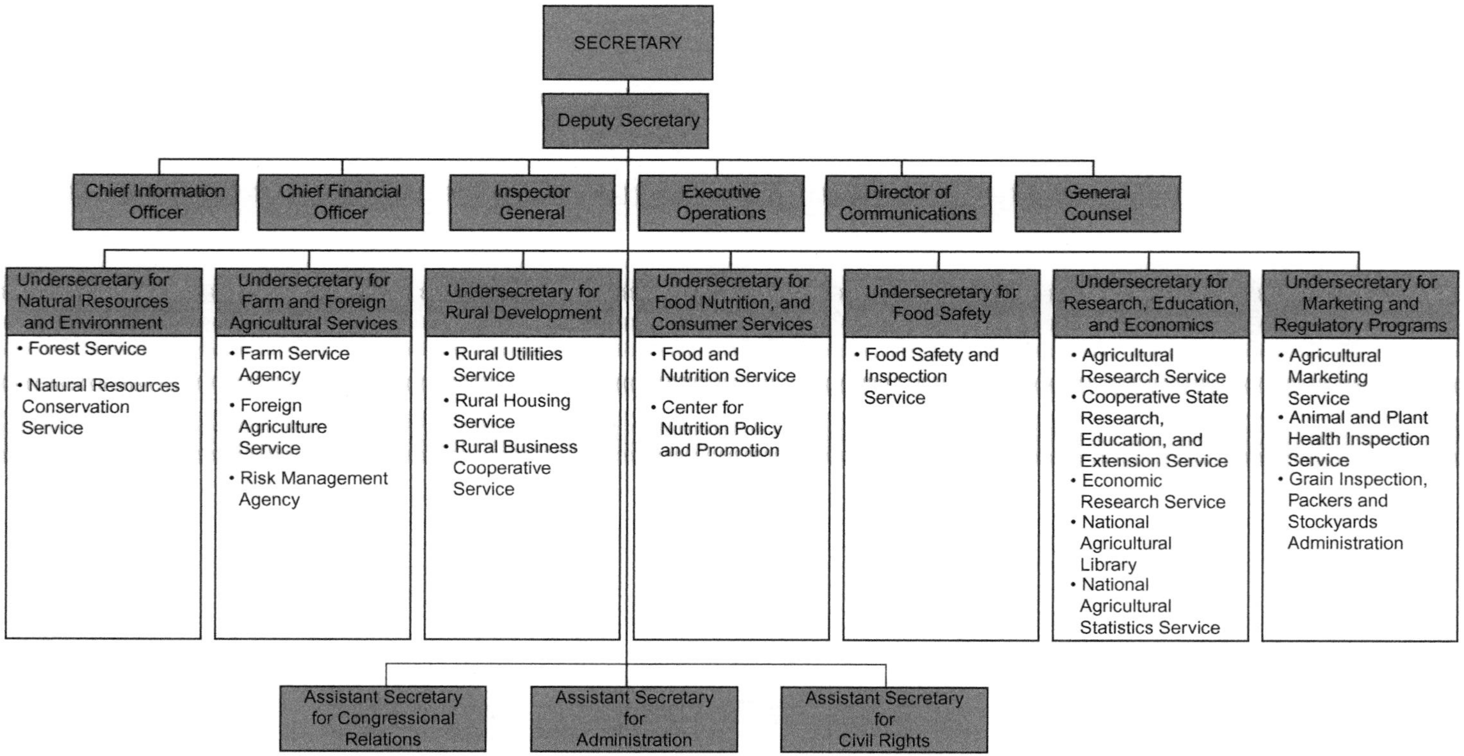

FIGURE 7.5 Organizational chart for the US Forest Service. *Credit: Encyclopedia of the Earth.*

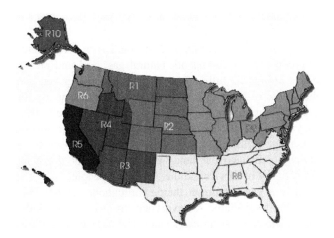

FIGURE 7.6 The nine US Forest Service regions. *Credit: US Forest Service. This figure is reproduced in color in the color plate section.*

other forest products are continuously available. Forest products include materials derived from a forest for commercial use, such as lumber and paper, and also "special forest products" such as medicinal herbs, fungi, edible fruits and nuts, and other natural products. The Office also maintains a silviculture inventory and tries to keep forest brush and prospective fuels down to reduce the potential of serious forest fires.

Each National Forest and Grassland is managed by a Superintendent. At a more local level, district rangers and their staff may be the first point of contact with the Forest Service. There are more than 600 ranger districts each with a staff of 10–100 people. The districts vary in size from 50,000 acres (20,000 ha) to more than 1 million acres (400,000 ha). Many on-the-ground activities occur in the ranger districts, including trail construction and maintenance, operation of campgrounds, and management of vegetation and wildlife habitat.

The US Forest Service hires wildlife biologists to watch over the immense numbers of game, non-game and species of concern on FS land. Other than federally endangered and threatened species, which often need special provisions, most of the FS efforts in managing wildlife occur through habitat management rather than population manipulation. The agency has several initiatives with catchy names such as "A Million Bucks" (deer), "Animal Inn" (protecting snags and dead trees as habitat) or "Dancers in the Forest" (grouse and ptarmigan) targeted at specific habitats or groups of wildlife.[5] They also team up with other organizations and the public in cross-cutting relationships such as PARC (Partners in Amphibian and Reptile Conservation), Partners in Flight (migratory birds) and Taking Wing (waterfowl). The FS is active in fisheries management, both anadromous and freshwater. It sponsors many programs in habitat assessment and improvement, population dynamics, and conservation of fisheries throughout the National Forest System.

More than 5 million acres (about 2 million ha) of forest lands are leased for energy production, including oil, natural gas, coal and phosphate. These leases produce around $2 billion of energy resources per year, and are managed by the Office of Minerals and Geology Management. By the way, panning for gold is permitted in most national forests; anyone is free to keep all the gold they can find! The Ecosystem Management Coordination Staff supports and manages planning and decision-making processes used by the Forest Service to manage the lands and resources of the National Forest System and delivery of services to the American people.

Education is a very common theme throughout the US Forest Service. Lifelong learning is available for everyone, from inner city kids to the millions of visitors to the forests each year. The Service provides free educational materials to schools and other youth organizations. The Kids in the Woods program and other educational opportunities reach nearly 4.5 million children per year.[6]

Unlike most of the agencies within the Department of Interior, the FS has active research programs under the Deputy Chief for Research and Development. The FS employs over 500 scientists to conduct research studies on forest, wildlife, fish, and range management; development of forest products; watershed management; urban natural resource issues; climate change; nanotechnology; biomass and bioenergy; fire management; outdoor recreation; and inventory and monitoring. The agency has 80 experimental forests and ranges that allow long-term research to occur. In addition, the FS maintains five Regional Research Centers, a National Forest Products Laboratory and an International Institute of Tropical Forestry. The Forest Products Laboratory, for example, studies a broad range of issues, from making safer baseball bats to using nanoparticles (particles with diameters of less than one-billionth of a meter and special properties due to their small size) derived from wood in various industrial processes. One of the earliest and still very important functions of the FS is to provide advice to states, tribes and private parties, including industry, regarding forest management; these are the functions of the State and Private Forestry Division. So, is this what they mean by multiple use?

A key component of forest management over the decades has been fire management. For most of the 20th century the FS had a strict policy of fire suppression. Any fire, caused by any reason including natural factors, was to be extinguished as quickly as possible. This policy was strongly influenced by huge, deadly fires. For example, the worst loss of life in United States history due to a wildfire occurred in 1871 when the Peshtigo Fire swept through Wisconsin, killing more than 1500 people.[7] The Santiago Canyon Fire of 1889 in California, and the Great Fire of 1910 in Montana and Idaho which killed 86 people, destroyed entire communities and burned 3 million acres (1.2 million ha), contributing to the philosophy that fire was a danger that needed to be totally suppressed. Therefore, the FS began a campaign throughout the nation to prevent and combat forest fires − and part of this campaign was the appearance of Smokey Bear in 1944 (Figure 7.7).

This adamant attitude against forest fires prevailed until the middle of the 20th century. Starting in the 1960s, foresters began to realize that total suppression of fire could also be detrimental because cessation of all fires allowed kindling and brush to build up on the forest floor, increasing the amount of available fuel if a fire should start and increasing the risk of a really serious fire. In concert with other agencies such as the National Park Service, the FS changed its policy from fire suppression to fire management in 1974. While human-caused fires were generally extinguished quickly, natural fires such as those started by lightning strikes were watched carefully but allowed to burn unless they threatened habitation or buildings. Prescribed burning was also allowed in some areas. In 2003 Congress passed *the Healthy Forests Restoration Act* to direct the FS to begin removal of built-up understory and brush and reduce the fuel load in National Forests. The Act recognized that accumulated brush provided a habitat for disease and harmful insects, in addition to promoting major fires. The Departments of the Interior and Agriculture have developed combined plans for fire management that establish uniform goals and methods of fire management. Even

FIGURE 7.7 The very first Smokey Bear poster, 1944.

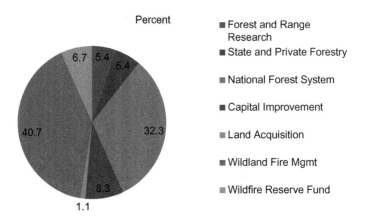

FIGURE 7.8 The budget of the US Forest Service based on percent of total budget. *Information from US Forest Service. This figure is reproduced in color in the color plate section.*

Smokey Bear has gotten into the Act — he used to say "Only you can prevent forest fires", but he now says "Only you can prevent wildfires".[8]

How Does the US Forest Service Spend Its Budget?

Somewhat surprisingly, the biggest outlay of the FS is not directly to the National Forest System itself but to fire management (Figure 7.8). Almost $2 billion of the $5.6 billion FS budget for 2012 went to fire-related programs. The National Forest System takes second place, with nearly a third of the budget. That leaves about 27% of the annual budget for all other programs.

NATURAL RESOURCES CONSERVATION SERVICE

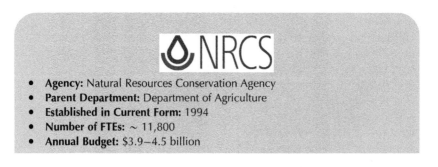

- **Agency:** Natural Resources Conservation Agency
- **Parent Department:** Department of Agriculture
- **Established in Current Form:** 1994
- **Number of FTEs:** ~ 11,800
- **Annual Budget:** $3.9–4.5 billion

- **Responsibilities:** Manages several conservation programs to help reduce soil erosion, enhance water supplies, improve water quality, increase wildlife habitat, and reduce damages caused by floods and other natural disasters. Prominent among these is the Conservation Reserve Program.
- **Important Laws or Acts:** National Industrial Recovery Act 1932, Soil Conservation Act 1935 (created Soil Conservation Service), Flood Control Act 1936, Standard State Soil Conservation Districts Law 1937, Watershed Protection and Flood Control Act 1954, Rural Development Act 1972, Resource Conservation Act 1985. Name changed to Natural Resources Conservation Agency 1994.

The chief responsibility of the Natural Resources Conservation Service (NRCS) is "to provide resources to farmers and landowners to aid them with conservation. Ensuring productive lands in harmony with a healthy environment is our priority" (from the NRCS motto[9]). They have staff specialists dealing with soils, water, crop depredation and wildlife management – essentially, the gamut of natural resource issues. The mission of the agency continues: "With operations in the United States, the Virgin Islands, Puerto Rico, and Guam, our agency touches the lives of a diverse range of individuals".[9] A very important part of the NRCS is to work with farmers in a variety of programs, including the very important Farm Bill and the Conservation Reserve Program.

History of the Natural Resources Conservation Service

I don't believe that very many of us under 80 years of age who are living in the United States today can appreciate the conditions of the Dust Bowl of the 1930s. Years of neglecting the soil, lack of fertilization, fall plowing exposing topsoil to the elements, and lack of cover finally caught up with the farmers of Oklahoma, Texas, New Mexico, Kansas and Colorado when a "perfect storm" of a severe drought hit the area. Soil turned to dust and, as winds whipped through the area, huge dust clouds formed, covering everything with a layer of silt (Figure 7.9). Over a two-day period in 1934, a huge dust storm carried some 350 million tons of silt all the way from the northern Great Plains to the eastern seaboard.[10] According to *The New York Times*, dust "lodged itself in the eyes and throats of weeping and coughing New Yorkers", and even ships some 300 miles offshore saw dust collect on their decks from the Midwest. More storms followed in 1935. No doubt, the Congressmen had to pay attention, as they dusted off their fedoras and suits.

In response, Congress passed *Public Law 74−46* in 1935,[11] which created the *Soil Conservation Service* (SCS) under the US Department of Agriculture. Congress recognized that "the wastage of soil and moisture resources of farm, grazing and forested lands ... is a menace to the national welfare". *Hugh Hammond Bennett (*Figure 7.10), considered the "Father of

FIGURE 7.9   A huge dust storm engulfs a farm in Oklahoma, 1935.

FIGURE 7.10   Hugh Hammond Bennett, the Father of Soil Conservation and first chief of the Soil Conservation Service. *Credit: Michigan Natural Resources Conservation Service.*

Soil Conservation", was instrumental in the formation of the agency and became its first Chief. Bennett became a crusader against soil erosion, and woke the country to the crisis that the nation was facing.

The SCS immediately set to work on soil erosion problems in the nation. Despite its name, the SCS also focused on watershed projects. The Flood Control Acts of 1936 and 1944 and the Agricultural Appropriations Act of 1953 provided funding for the SCS to work on these projects. Since 1944, the SCS (now NRCS) has built almost 11,000 dams on approximately 2000 watersheds for flood control, recreation, and water supplies.[11]

From its inception, the staff in the SCS realized that soil erosion could only be controlled by working with the private farmer because over 70% of the land across the country is in private ownership. To organize this effort, the SCS formed watershed-based *Soil Conservation Districts*. From the first Soil Conservation District in North Carolina the program has grown to more than 3000 districts nationwide.

Later, in the 1950s, the Service provided assistance on multiple conservation issues in the Great Plains, and to the *Soil Bank Program* which paid farmers to set aside highly erodible lands from production. During the 1960s the agency expanded into both rural and urban programs. The *Resource Conservation and Development Program*[12] facilitated working in larger areas than watersheds for long-term economic development of rural areas. As suburban areas developed from previous farmlands, the SCS provided guidance on avoiding erosion and developing recreational opportunities.

During the Ecology Movement in the late 1960s and 1970s, the SCS responded to the increased awareness of the American people by focusing on water quality issues. It also gained authority to monitor the natural resource base of the United States under the *National Resources Inventory* in the Rural Development Act of 1972.[13] In 1977, the Soil and Water Conservation Act mandated that the US Department of Agriculture regularly report on the condition of water and soil resources on non-federal lands, and this responsibility was handed down to the SCS.

In the 1980s, a farm crisis caused catastrophic economic losses in the nation.[14] Its causes were similar to those of the housing problem or "bubble" 30 years later. Farm production was high in the late 1970s and early 1980s, but so was international demand for crops. Increased demand and lots of product to sell led to substantial farm profits. Farmers used the promise of future incomes to purchase more land by taking out loans, and this led to increased land values; everything seemed to be booming. However, economic cycles eventually return to their starting point, and after a few years commodity prices started dropping. Farmers were not able to meet their mortgages, causing land values to plummet, and many farmers found that their debts exceeded their total assets. To help support farms, the government, through the USDA, established programs to deal with this crisis, and the SCS took the opportunity to implement several conservation programs, including the *Sodbuster*, *Swampbuster* and *Conservation Reserve Program*. These programs will be discussed in more detail below. Additional conservation measures, which will also be discussed below, were enacted in the 1990s. In response to these new programs and a significantly increased set of mandates, the SCS was renamed the Natural Resources Conservation Service (NRCS) in 1994. Since then, the NRCS has been heavily involved with some major conservation programs.

Organization of the Natural Resources Conservation Service

The bureaucracy of the NRCS has the usual sectors for business operations, personnel, finances, and offices for interacting with the Congress and the public found in virtually all agencies. In addition, there are four regional offices that oversee operations within different parts of the nation, and four Deputy Chiefs (Figure 7.11). The duties and offices of the Deputy Chiefs include the following.

1. *Science and Technology.* This directorate leads NRCS efforts for science-based technology in conservation engineering, air quality, energy conservation, nutrient management technology, and other ecological issues. It has responsibility for the NRCS Conservation Innovation Grants Program, which funds proposals from state and non-governmental agencies, tribes and individuals for the development and implementation of new ideas concerning conservation of natural resources. The Plant Materials Center under this directorate provides all sorts of information concerning plants in the United States; it maintains the online Plants Database[15] so that anyone can find information on all sorts of vegetation. The Science and Technology Office collaborates with the Soil Survey and Resource Assessment area to set NRCS research and technology development priorities; with other federal agencies, tribal, state, and local governments; with academia seeking best-science technologies and technical tools for natural resources conservation; and with foreign nations to provide technical support on conservation issues.

2. *Programs.* This area is responsible for handling a variety of assistance programs, easements and financial technical assistance. This office reimburses landowners for approved conservation measures conducted on their lands, such as wildlife habitat improvement and wetland restoration. Payments for land set-asides under the Conservation Reserve Program, however, come through the Farm Service Agency.

3. *Soil Survey and Resource Assessment.* This directorate supervises Divisions, Institutes and Centers related to soil science, soil survey and resource assessment, technical assistance to foreign governments, and international scientific and technical exchange. It has technical leadership for the use of geospatial technologies (GIS, GPS, remote sensing) in NRCS and for establishing geospatial data standards for the agency. The Soil Survey section has four centers concerning soils, water, climate, and geospatial information. These centers primarily provide help to other agencies, standardize methodology, and provide data to users.

4. *Strategic Planning and Accountability.* This area primarily serves the internal programs of the NRCS, rather than outreach like the other three areas.

This organization helps the NRCS to function but if you are a landowner or farmer you probably know the NRCS best through its extension services. Each state has a service center that coordinates conservation efforts for the entire state. However, most counties throughout the nation also have one or more local service centers that house staff of the NRCS for technical

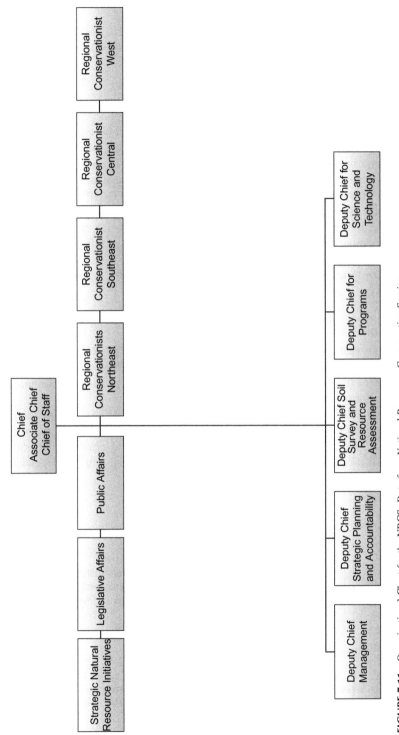

FIGURE 7.11 Organizational Chart for the NRCS. *Data from National Resources Conservation Service.*

assistance and the Farm Services Agency for financial assistance, and an office for the local conservation district. Through these offices, the USDA provides direct one-to-one help to landowners.

The NRCS Budget

About 20% of the NRCS annual budget of around $4.5 billion is discretionary, with the rest being obligated to existing programs (Figure 7.12). Almost all of the mandatory portion goes to reimbursement programs for land and wildlife conservation. For example, over 38% of the mandatory budget goes to the Environmental Quality Incentive Program (EQUIP), 21% to the Wetlands Reserve Program (WRP) and 21% to Conservation Stewardship. The Farm Services Administration covers the budget for the well-known Conservation Reserve Program, so it is not included in this budget. These programs are described in greater detail below. Of the discretionary portion of the budget, 75% to 85% is spent on providing technical assistance in various ways.

Farm Bill Conservation Measures

As mentioned above, one of the primary functions of the NRCS is to provide technical assistance on soil, water and farm wildlife. However, a huge fraction, more than three-quarters of its budget, is used to provide this technical support for programs mandated by Farm Bills.

Farm Bills are vast Congressional commitments to US agriculture, commodity price supports, crop and disaster insurance programs, farm loans, nutrition programs, international trade agreements, rural development, rural electrification programs, agriculturally based research, aid to dependent

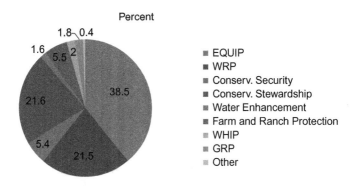

FIGURE 7.12 The annual mandatory budget for NRCS. The total budget for NRCS approximates $4.5 billion, but the mandatory portion, seen here, is around 80% of that, or approximately $3.6 billion. *Information from US Department of Agriculture. This figure is reproduced in color in the color plate section.*

children, food stamps and many other elements. Sometimes huge bills such as the Farm Bill are called *Omnibus Bills* because of their complexity. The 2008 Farm Bill (formally known as the Food, Conservation and Energy Act of 2008) obligated $288 billion over a five-year period.

Over the past several years, Farm Bills have included some very wide-ranging and important conservation measures. The NRCS is the principal advisor in these projects and finances some of the programs through the Deputy area for Programs, while the Farm Services Agency finances other conservation projects. There are two broad categories of conservation programs in this package: voluntary programs that provide financial assistance to growers for doing something positive, and mandatory ones that penalize landowners for harmful activities.

Some of the major voluntary conservation programs that provide financial and technical assistance to growers include the following.

1. *Environmental Quality Incentives Program* (EQUIP).[16] This program provides technical and financial assistance to farmers to plan and implement many kinds of conservation projects, including improvement of soil, water, plant, animal, air and related resources, on agricultural land and non-industrial, private, forestland. The criteria for receiving financial aid come from each State office of the NRCS — not everyone who applies is eligible. Financial aid is on a cost-share basis, where both the NRCS and the farmer contribute. In the fiscal year 2011 the program spent more than $864 million and included more than 13 million acres (5.3 million ha) nationally.[17] In terms of total acres and total budget, EQUIP is the largest conservation program funded by the NRCS.

2. *Wetlands Reserve Program* (WRP).[18] In some states, 95% to 99% of the original wetlands have been drained and cultivated. In the conterminous United States, approximately 48% of wetlands that existed in the 1600s have been lost.[19] Wetlands provide many ecological services, such as flood control, water storage, enhanced species diversity, and habitat for fish and wildlife. The Wetlands Reserve Program provides assistance to landowners to protect, restore and enhance wetlands on their property. Landowners who contract their land into permanent or 30-year easements receive payments from the NRCS and up to 100% of the costs to restore the wetlands. Between 1992 and 2011, the NRCS signed more than 13,000 contracts and restored almost 2.5 million acres (1 million ha) of wetlands.

3. *Wildlife Habitat Incentives Program* (WHIP).[20] This is a voluntary program for landowners who wish to improve their lands for fish and wildlife populations. The NRCS provides up to 90% of the costs of approved projects. Specific objectives of the program include: (i) to promote the restoration of declining or important native fish and wildlife habitats; (ii) to protect, restore, develop or enhance fish and wildlife habitat to benefit at-risk species; and (iii) reduce the impacts of invasive species on fish

and wildlife habitats; protect, restore, develop or enhance declining or important aquatic wildlife species' habitats; and protect, restore, develop or enhance important migration and other movement corridors for wildlife. Special attention is given to enhancing the habitat for species that are endangered, threatened or of special concern. In a given year, the NRCS enacts around 3800 contracts covering around 850,000 acres (344,000 ha) and costing approximately $60 million.

4. *Conservation Reserve Program* (CRP).[21] The Conservation Reserve Program is a major program focusing on highly erodible soils. For this program the Farm Services Agency, through its Commodity Credit Corporation, provides the finances but the NRCS provides technical assistance. This is a voluntary program for agricultural landowners through which they can receive annual rental payments and cost-share assistance to establish long-term, resource conserving covers on eligible farmland. Participants enroll in CRP contracts, usually for 10 or 15 years, and receive payments from the government each year of enrollment. However, lands are required to meet certain criteria to be enrolled: (i) they must be held by the enrolling owner for at least a year; (ii) they must have been in production during four of the previous six years before their first enrollment; (iii) they must be highly erodible (Figure 7.13); and (iv) they must be in a CRP conservation priority area. Lands approaching the end of their enrollment period may be re-enrolled. Enrolled lands are initially planted in cover crops such as grasses to protect the soil and promote soil binding. However, in 2003 the USDA determined that lands in certain types of CRP had to be subjected to *Mid-Contract Management*[22] to enhance the land for wildlife. Much of the CRP in the Midwest, for example, had grown into rank, monotypic stands of grass that provided little wildlife value. On a 50% cost share with the FSA, farmers can use a variety of techniques to break open the habitat, induce more species of plants, and encourage grassland song birds and game birds.

The CRP's regulations dictate that some lands may not be enrolled. Each state is given a maximum amount of land it can enroll in various CRP programs. The state then determines how many acres it wants to enroll in the CRP program for a specific sign-up period, and the amount of funding it will provide per acre. Sign-ups generally occur twice per year, but some years have had only one sign-up since the program began in 1986. Each application is evaluated by NRCS staff according to an environmental benefits index. Contracts are offered in order of the index until the upper quota has been reached. In a recent sign-up, for example, 14% of the lands offered nationally were rejected.[23] The size of the program is impressive – about 714,000 contracts on 416,000 farms, encompassing 31.3 million acres (12.7 million ha); this is about the size of the state of North Carolina. Annual rental payments average about $53/acre for a total yearly cost of $1.7 billion.[24]

FIGURE 7.13   An example of very erodible land planted into corn. *Credit: Center for Rural Affairs.*

Other, smaller programs provide landowners with technical and financial assistance on conservation stewardship, water programs, grasslands and forests.

So, what is the value of the Conservation Reserve Program to natural resources? The first value, of course, is to remove erodible land and protect it from further erosion, thus the program is instrumental in soil conservation. Over the years, over 329 million acres (121 million ha) have been conserved in all programs (Figure 7.14). Because lands are protected or improved there is less wind erosion, which provides a benefit to clean air and water. A second value, derived by the matching cost supports of the program, is enhancement of wild-life habitat. Mid-contract management, initial planting into native grasses, development of buffer and filter strips around field edges and between fields and waterways, and planting some areas into forests instead of grasslands, are measures that greatly assist the conservation of our natural resources.

There are two major mandatory programs that penalize farmers for dam-aging the environment — Swampbuster and Sodbuster. Both of these were enacted with the 1985 Farm Bill and have been renewed ever since. The dif-ference between these programs and those above is the approach. Swampbuster and Sodbuster provide penalties for violations, whereas CRP, EQUIP and the others provide support — they *give* landowners money to implement conservation practices. It is not difficult to guess which of the two methods has more appeal.

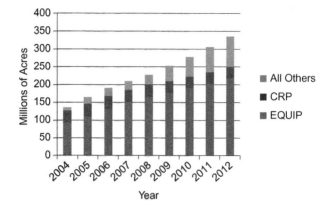

FIGURE 7.14 Accumulated number of acres conserved by NRCS and CRP programs, 2004–2012. *Data from US Department of Agriculture.*

Swampbuster arose because the NRCS and other agencies are deeply concerned about the loss of wetlands in the United States. Between the mid-1950s and 1970s, about 87% of wetland loss was due to agricultural practices. During dry periods many farmers lay field tiles in wetter portions of their fields to keep them drained. Many of these wet fields are actually natural temporary to ephemeral wetlands used by migrating waterfowl, and provide the ecological services mentioned above when they are most needed. To combat this, Congress enacted the Swampbuster Act into the Farm Bill. According to this Act, farmers who plant an agricultural commodity on wetlands that were drained after December 23, 1985 are ineligible for program benefits in any year a crop is planted. Potentially lost benefits include all USDA farm programs including subsidies and price supports, and any cost-sharing of the NRCS programs mentioned above.

Sodbuster was established in 1984 because of continued soil erosion problems. The Act provides disincentives to farmers and ranchers who produce annually tilled agricultural crops on highly erodible cropland without adequate erosion protection. It operates similarly to the Swampbuster Act, in that farmers who till highly erodible lands are subject to a loss of their USDA program supports.

Study Questions

7.1. Conduct a review of the Dust Bowl Era. What led up to the poor soil conditions that interacted with climate to cause this environmental disaster? Were there any effects of the Dust Bowl Era in your state — either directly or indirectly (e.g., through immigration of farmers)?

7.2. Are you at a land grant university, or are you familiar with the land grant institutions in your state? Name some distinguishing features of land grant universities compared to non-land grant institutions that exist even to today.

7.3. Was there any correspondence between the history of wildlife management and forest management in the United States in regard to specific periods and concerns?
7.4. What do Aldo Leopold, Gifford Pinchot and Hugh Hammond Bennett have in common?
7.5. Trace the history of the US Forest Service from its inception to its present condition.
7.6. Do you, or does anyone you know, have land that is enrolled in one of the NRCS conservation programs? What do you/they think of the program? Is it worthwhile?

REFERENCES

1. US Department of Agriculture. Mission Statement. <www.usda.gov/wps/portal/usda/usda-home?navid=MISSION_STATEMENT>; 2013.
2. US Forest Service. Mission Statement. <www.fs.fed.us/aboutus/missions.html>; 2013.
3. US Forest Service. About us. <www.fs.fed.us/forestmanagement/aboutus/histperspectives.html>; 2013.
4. Conrad D. *The Land We Cared for ... A History of the Forest Service's Eastern Region.* US Forest Service. George Banzoff Co., Milwaukee, WI, for the US Forest Service; 1997.
5. US Forest Service. Wildlife. <http://www.fs.fed.us/biology/wildlife/index.html>; 1999.
6. US Forest Service. Conservation Education. <http://www.fs.usda.gov/main/conservation-education/home>.
7. US Forest Service. About us. <http://www.fs.fed.us/fire/people/aboutus.html>.
8. US Forest Service. Only you can prevent wild fires. <http://www.smokeybear.com/wildfires.asp>.
9. National Resource Conservation Service. People. <http://www.nrcs.usda.gov/wps/portal/nrcs/main/national/people>>.
10. History. <www.history.com>; 2013.
11. Natural Resources Conservation Service. <www.nrcs.usda.gov>.
12. National Association of Resource Conservation and Development Councils. <http://narcdc.org/>.
13. Federal Register 42 USC. 3122, <http://www.csrees.usda.gov/about/offices/legis/pdfs/rural devact.pdf>.
14. Nebraska Studies.org. <www.Nebraskastudies.org>.
15. US Department of Agriculture. Natural Resource Conservation Service. Plants Database. <http://plants.usda.gov/java/>; 2013.
16. Natural Resources Conservation Service. Environmental Quality Incentives Program. <http://www.nrcs.usda.gov/wps/portal/nrcs/main/national/programs/financial/eqip/>.
17. US Department of Agriculture. Farm Services Agency. <http://www.fsa.usda.gov>.
18. Natural Resources Conservation Services. Wetland Reserves Program. <http://www.nrcs.usda.gov/wps/portal/nrcs/main/national/programs/easements/wetlands/>.
19. US Environmental Protection Agency. Wetlands – Status and Trends. <http://water.epa.gov/type/wetlands/vital_status.cfm>; 2012.
20. Natural Resources Conservation Service. Wildlife Habitat Incentives Program. <http://www.nrcs.usda.gov/wps/portal/nrcs/main/national/programs/financial/whip/>.

21. Natural Resources Conservation Program. Conservation Reserve Program. <https://www.fsa.usda.gov/FSA/webapp?area=home&subject=copr&topic=crp>; 2013.
22. Farm Services Agency. <www.fsa.usda.gov>.
23. Farm Services Agency. Conservation Programs. <http://www.fsa.usda.gov/FSA/webapp?area=home&subject=copr&topic=crp-st>; 2013.
24. Farm Services Agency. <http://www.fsa.usda.gov/FSA>.

Chapter 8

Natural Resource Management at the State and Provincial Levels

Terms to Know

- 10th Amendment to the US Constitution
- Constitutional Monarchy
- Crown lands
- Commissions
- Federal Aid for Wildlife Restoration Act
- Federal Aid for Sport Fishing Restoration
- State Wildlife Grant

STATE AND PROVINCIAL AUTHORITY

Individual states and provinces have considerable power in determining how the resources on their lands are to be managed. In the United States, the *10th Amendment to the Constitution* states "The powers not delegated to the United States by the Constitution, nor prohibited by it to the States, are reserved to the States respectively, or to the people". As we saw in Chapter 7, the United States Constitution specifically states that the federal government has jurisdiction over treaties with other nations, interstate commerce, and protection of federal property including lands. In 1973, endangered species also became a federal responsibility shared with the states. The Canadian Constitution clearly distinguishes the authority of the provinces and the federal government, and gives the provinces primary responsibility for natural resources within their borders. For most of these resources, the provinces seem to have greater delegated power than in the United States (see below).

In both the United States and Canada the individual states and provinces follow a structure that is miniaturization of their respective federal governments. In the United States, most states have bicameral legislatures consisting

D.W. Sparling: Natural Resources Administration. DOI: http://dx.doi.org/10.1016/B978-0-12-404647-4.00008-8
199

of a larger House of Representatives (also called Assembly or House of Delegates) and a smaller Senate. The legislative branch of each state in general determines the overall responsibilities and capabilities of agencies within the executive branch, but leaves the specific details of implementing the functions to the executive branch. For example, a law or Act passed by a state legislature may determine that the state's Department of Conservation shall regulate the harvest of non-migratory game species within the state consistent with maintaining healthy populations. The legislature may further determine which species are game and which are not. Thus, the legislature establishes major directions. However, the nuts and bolts of actual management, including establishing seasons, bag limits, harvestable sexes and suchlike are left up to the agency.

The governor is the chief executive officer of the state, is elected in a general election, and subsequently establishes operational policy in consultation with the agency directors. The governor also has to sign all bills issued by the legislature before they become law, and thus has final approval or veto power. In most states, the directors of the various agencies collectively constitute the state cabinet. These directors may be appointed by the governor (often with legislature approval). Alternatively, and in contrast to the federal government, some states have commissions that hire and fire the directors (see below). Unless they are impeached by the legislature, the governor and the state cabinet serve for a defined term after which elections are held again, such as every four years.

In addition to the states, the United States has nine territories,[1] five of which are inhabited. Most of these are islands in the Pacific or Caribbean. The inhabited islands have non-voting delegates to US Congress, local voting rights, and are protected by the federal military. Together they include about 4 million Americans, 91% of which are in Puerto Rico. Except for endangered species, the territories are free to manage their natural resources as their populations and governments see fit.

Canadian provincial governments have only one legislative arm, the *Legislative Assembly*. Members of this assembly are called Members of the Provincial Parliament (MPPs) in Ontario and Members of the Legislative Assembly (MLA) in the other provinces; they are elected by the people. As in the federal government, members are appointed to various department committees by the head of provincial government and to the office of minister for each department. The head of the provincial government, or Premier, is also a member of the legislature and is appointed by the Lieutenant Governor, representing the Crown. Thus, in parliamentary government there is no separation of branches as seen in the democratic structure of the United States.

Canada also has both provinces and territories, but the territories are contiguous with the provinces and are much larger than in the United States (Figure 8.1). Territories are in northern Canada, and, while they have 40% of the nation's land mass, they only have 3% of the population,[2] most if which

FIGURE 8.1 Canada is composed of provinces and very large territories in its northern reaches. *Credit: Environment Canada.*

is Inuit or First Nations. Territories fall within the jurisdiction of the federal government.[3]

An important aspect of provinces and territories is the *Crown lands*, which can be either federal or provincial. Crown lands consist of real property that belongs to the monarchy and must be passed down from one generation to another and cannot be sold.[4] In practice, they are public lands held in perpetuity and contain extensive and diverse natural resources. About 41% of the land mass of Canada is held as federal Crown lands (mostly in the territories), 48% as provincial Crown lands and 11% as private lands.[4] The percent of province in Crown lands varies from 95% in Newfoundland and Labrador, and 94% in British Columbia, to 2% on Prince Edward Island. Close to 98−99% of the territories comprise federal Crown lands.

SO WHAT DO STATES AND PROVINCES MANAGE?

States set regulations for recreational and commercial use of resources on non-federal lands, manage real estate owned by the state, establish harvest regulations for fish and wildlife, set laws on mineral extraction, and enforce regulations concerning the use of pesticides and other agricultural chemicals. In contrast, provinces may set laws on hunting and recreational fishing but cannot regulate commercial fishing, even in their own freshwaters. Most of these government units also regulate pollution within their geographical jurisdictions, although usually the federal government adjudicates atmospheric, marine or water pollution. Every province and state has parks, forests and wildlife areas. In both the United States and Canada the federal government takes the lead in identifying and conserving species at risk (endangered or threatened in the US), but provinces and states can declare their own lists for species within their borders. States that border oceans have limited authority over recreational use of marine resources up to three nautical miles from their coasts (nine nautical miles for some states along the Gulf Coast[5]). Inland boat registration is a part of state authority. Provinces can only regulate coastline use; the federal government oversees all regulations at sea. The licensing of boats and boating training courses are also conducted by the Canadian federal government.

As an example of state authority over wildlife, take a look at Table 8.1; this is a condensed copy of an actual published list of regulations concerning game laws. Other states and provinces may vary in detail but, in general, have similar regulations. States typically classify the type of game into categories such as those in the table. Harvest limits may be set based on sex when the sex of a species can be easily determined in the field and when one sex contributes more to the population per average individual than does the other sex. For example, male white-tailed deer (*Odocoileus virginianus*) with antlers are readily distinguished from does. Moreover, since only a fraction of bucks but almost all does mate in any given year, each doe has a higher probability of contributing to the next generation than an average buck, and differential harvest limits may be set depending on whether the state wants to increase or decrease its herd of deer. Because yearling bucks may have very small antlers or "buttons" during their first hunting season, actual regulations may specify antlered versus *antlerless deer* instead of males and females.

Each species also has open and closed seasons, based on the biology of the animal. Open seasons usually occur in autumn, when few animals are giving birth. Hunting times may also be based on a species' behavior patterns or during safer times to hunt. Agencies also distinguish the type of weapon that may be used – traps, rifle, shotgun, muzzleloader, bow and crossbow, with further regard to types of trap, caliber, shot size and bore, and arrow release speed. Further, states declare the harvest limits, with daily limits being the number of animals that can be harvested on a given day per license and possession limits being the number of animals a person can have in possession, including storage, per license. There are other regulations concerning hunter education, firearm registration and license or permit fees that

TABLE 8.1 A Summary of a State's Harvest Regulations.[2]

	Species	Dates and zones	Hours	Daily limit	Possession limit
Small Game	Rabbit		Sunrise to Sunset	4	10
	Cock Pheasant			2	6
	Hungarian Partridge			2	6
	Quail			8	20
	Squirrel		Half an hour before sunrise to half an hour after sunrise	5	10
	Woodchuck			No limit	
Deer and Turkey	Firearm Deer	Nov 16–18	Half an hour before sunrise to half an hour after sunset	1 deer per firearm permit	
	Muzzleloader Deer	Dec 7–9		1 deer per muzzle-loading permit	
	Special CWD Season	Dec 27–30		1 deer per valid permit	
	Late Winter Antlerless Deer			1 antlerless deer per permit	
	Deer Archery in Selected Counties	Oct 1–Nov 15		1 deer per archery permit	
	Deer Archery in Selected Counties	Oct 1–Jan 20			
	Youth Firearm Deer	Oct 6–7		1 deer	
	Youth Turkey	Mar 30–31	Half an hour before sunrise to 1 pm	1 gobbler or bearded hen	
	Turkey Spring	Apr 8–May 9		1 gobbler or bearded hen per permit	
	Turkey Fall Shotgun	Oct 20–Oct 28	Half an hour before sunrise to sunset	1 either sex turkey per permit, maximum of 2	
	Turkey Fall Archery	Oct 1–Jan 20	Half an hour before sunrise to half an hour after sunset		

(Continued)

TABLE 8.1 (Continued)

	Species		Dates and zones	Hours	Daily limit	Possession limit
Migratory Game Birds		Dove	Sep 1–Oct 28	Sunrise to sunset	15	30
		Teal	Sep 8–23	Half an hour before sunrise to sunset	4	8
		Early Canada Goose	Sep 1–15		5	10
		Rail	Sep 8–Nov 16	Sunrise to sunset	25	25
		Common Snipe	Sep 8–Dec 23		8	16
		Woodcock	Oct 20–Dec 3		3	6
		Crow	Oct 28–Feb 28	Half an hour before sunrise to sunset	No limit	
Furbearers	Trap	Raccoon	Nov 5–Jan 20	Unrestricted	No limit	
		Fox	Nov 10–Jan 31	Unrestricted		
	Hunt	Coyote	Year round	Half an hour before sunrise to half an hour after sunset		
		Raccoon	Nov 5–Jan 20		None	
		Beaver	Nov 5–Mar 31		None	
		River Otter	Nov 5–Mar 31		5	

apply to every hunter. All of these regulations are established by the agency in following its mandate from the state legislature to manage populations of game animals or in the legislation promulgated by the ministry.

A huge responsibility of both state and provincial agencies is the management of people (see Chapter 9). Facility managers tend to agree that they spend more time managing people than wildlife. A big part of their job is making sure that people are safe, staying in areas that keep them from getting hurt, and following regulations that facilitate a safe and enjoyable environment. Managers may also have to keep the public away from sensitive areas, such as around the nests of Bald Eagles (*Haliaeetus leucocephalus*) or bear dens.

ORGANIZATION AT THE STATE LEVEL

In administering these natural resources, states seem to relish developing their own unique organizational structures. To some degree this is undoubtedly due to differences in the history of the state and to the individuality of leading politicians.

State natural resource administrations, just as in federal government, have several offices. Some are directly involved with the natural resources they represent; others run the business of the agency. A generalized and highly simplified organizational chart can be divided into three broad segments of administration, resource, and technical or business, headed by a director who may or may not be responsible to a commission that is appointed by the governor (Figure 8.2).

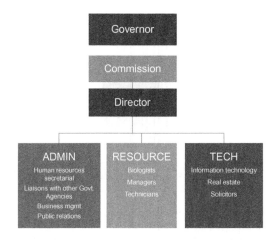

FIGURE 8.2   State agencies typically include a section for running the operations of the agency, another one for technical details such as information transfer, and a third section that works directly with natural resources. This is a generalized organizational chart for a state wildlife agency.

There are many different variations of this generalized chart. In a few states, each natural resource (forests, parks, fish, wildlife, marine, minerals, fossil fuels) is managed by a separate agency (Figure 8.3). On the one hand, this system might lead to duplication of effort and cost more to operate than some combination of agencies. On the other hand, it is likely that each resource has a better chance of receiving the attention it deserves than if agencies are combined. In still other states, we find a natural resources department that covers some resources but may exclude fisheries, wildlife, forestry, mining, or a combination of these. One of the most common combinations, in

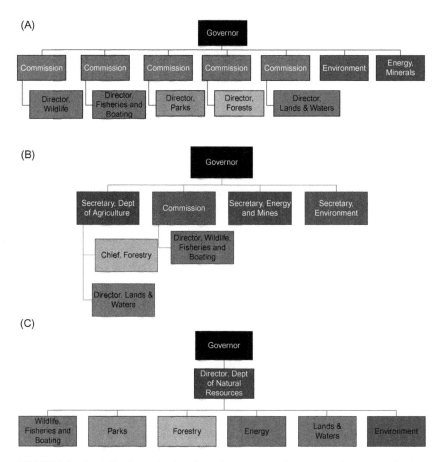

FIGURE 8.3   Generalized organization charts for state natural resources. (A) An organization with each agency existing as a separate entity. (B) A state where there is some consolidation of agencies. Regulatory agencies such as Environment and Energy and Minerals typically have no commission over them. (C) A state with an umbrella Department of Natural Resources containing the separate resource agencies. In these examples, the business and technical components are not represented.

about 11 states, includes fisheries, wildlife, boating, and conservation law enforcement (Figure 8.3B). The next most common grouping includes the above plus parks, while other states toss in forestry, or coastal and marine if they border an ocean. Fifteen states have other agencies in various combinations, but keep forestry separate either under a Department of Agriculture or as a separate forestry commission; three states house their forestry management within state universities. Many states separate their natural resource management agencies (e.g., fish, wildlife, parks) from their environmental protection agencies and mineral boards (i.e., those that have more of a regulatory than a management function). In general, there has been a gradual trend to centralizing the administration of natural resources. Twenty-five years ago, only four states had combined departments of natural resources that contained all of the divisions over the various resource types (Figure 8.3C)[6]; in 2012, nineteen states had combined departments (see Appendix A to this book).

Common inclusions among state natural resource agencies are *commissions* or boards. Of the 115 agencies listed in Appendix A, 61 have commissions, although the functions of the commissions vary considerably. Typically, the commissioners are appointed by the governor. Ideally, commissioners should be people experienced in natural resource management, or at least have business acumen. Unfortunately, filling these slots with patronage appointees is not unheard of.

Some commissions or committees have very specific advisory functions, such as endangered species or land zoning; these are not included in the Appendix. Among commissions that have broader responsibility for a particular agency, about 75% are principally advisory. They help to establish policy, represent various districts within the state, and even hire or appoint the chief executive officer. These commissions delegate the responsibility and most of the authority of running the daily activities of the agency to the director. The other 25% of commissions are more supervisory, and serve as a board under a chairperson to run the agency. In other words, the commission functions as both the advisory group and the director. Few of the comprehensive natural resource departments have commissions, and, if they do, the commissions are strictly advisory. Among the 20 stand-alone fish and wildlife agencies listed in Appendix A, all have commissions. In contrast, only a third of the stand-alone park agencies and about half of the forestry agencies have commissions.

In 2010, the Association of Fish and Wildlife Agencies conducted a survey of state fish and wildlife agencies regarding the structure and function of boards and commissions.[7] Some of their findings included the following:

- 67% of the states have the governor appoint the commission members
- Virtually all commission members are volunteers, with compensation including only travel costs (if that)
- 34% hire and fire the agency directors

- 80% elect their own chairs; in other cases the governor appoints the chair
- About 75% of the state agencies are served by only one board, but some may have up to four boards
- 61% of the commissions have 6 to 10 members, but there can be as many as 16
- Term lengths tend to be four years or more, and most agencies do not have limits on the number of terms a person can hold.

As with virtually everything else in state natural resource management, the nature of the CEO's position and title varies. "Director" or "commissioner" (even with agencies that do not have commissions) are the most common titles, but other labels include executive director, secretary, and state forester (for, obviously, forestry agencies). Regardless of the title, the primary function of the CEO is to manage the agency and oversee all of its functions. If the commission establishes policy, it is up to the CEO to implement that policy. The job really is similar to that of a top executive in a company with a board of directors, and typically does not involve much, if any, actual hands-on with natural resources. It's a truism that the higher a person gets in an agency, the less direct official involvement that person usually has with nature. That's not to say that directors do not get out, but if they do it is often more for recreation than for work. The CEOs of these agencies find their job through several venues. About 70% are hired by the governor, often with approval from the legislature. Another 24% are employed by the commission, with approval from the governor or the legislature, and around 6% are elected by the people of the state.

STRUCTURE WITHIN PROVINCES

Given that there are only 10 Canadian provinces, they show almost the same range in the numbers and kinds of agencies as do states in the United States (Appendix B). Most have some sort of Ministry (or Department) of Natural Resources that houses many of the management agencies, and a Ministry of the Environment that most often includes the chief regulatory offices. Other offices for energy, mines and minerals are also prevalent. As examples, let us take a look at a couple of organizational charts. Figure 8.4 shows the chart for the Manitoba Ministry of Conservation and Water Stewardship. In this case, the chief administrators – the minister and his deputy – divide the responsibilities for leading the ministry. The minister oversees several boards and advisory committees. These committees advise the minister on the various activities within the ministry, and help establish policy. The jurisdiction of most of these committees and boards is self-explanatory, but further information can be gained from the ministry's website. The primary job of the deputy minister (DM) is to oversee the operations of the ministry. Because much of the natural resources are on Crown lands, the DM is guided by

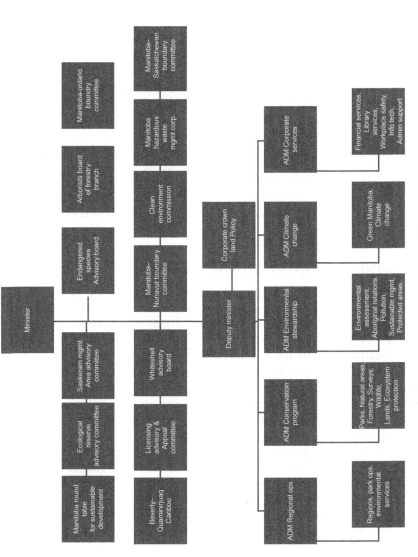

FIGURE 8.4 Partial organizational chart for the Manitoba Ministry of Conservation. Each of the Assistant Deputy Ministers oversees several other offices, which are not on this chart due to lack of space. See text for explanation. *Credit: Province of Manitoba.*

Corporate Crown Land Policy. Assistant deputy ministers (ADMs) supervise specific elements within the ministry, including regions, parks and other lands, pollution prevention and environmental assessment, climate change concerns, and business operations.

As mentioned, Ministries of the Environment typically are the primary environmental regulators for the provinces. The Ontario Ministry of the Environment (Figure 8.5), for example, deals with water, air and land pollution, waste management, and environmental monitoring. Its head is composed of the minister and his deputy, and, because they share the same organizational box, the chart suggests that they share responsibilities rather than divide them. The minister and his deputy receive input and advice from various advisory committees on water and pesticides. They also oversee real estate, legal services, finances and communications. ADMs oversee specific operations, including drinking water, business operations, regions, environmental monitoring through both field and laboratory operations, implementation of policy, and other environmental affairs such as working with First Nations. Keep in mind that the provinces have the chief responsibility for most aspects of the environment, and these complex organizations are consistent with those duties.

FUNDING FOR STATE NATURAL RESOURCES

It takes quite a bit of financing to manage natural resources. The primary sources of funding for state agencies include general funds from the legislature, permits and licenses, grants from the federal government, interest income on revenues, and miscellaneous funds. Park departments in many states charge user fees for camping and other activities. Forestry agencies may obtain funding from the sale of timber harvests or the collection of firewood. Agencies that supervise minerals and mines often make profits on extraction fees, or by issuing permits to mining companies. Environmental protection agencies place fines on polluters, which at times can yield high rates of return. The allocation of these funds varies – in some cases the agency is allowed to keep all or a part of what it collects; in other cases the fines and fees go the general funds of the state. The state general assembly may also allocate funds for particular purposes associated with natural resources, including pollution prevention or clean-up, support of state parks, monitoring, and other activities. The proportion of state funds spent in this way varies considerably, from slightly over 1% to perhaps more than 10% in some states. Let us take a closer look at the funding of fish and game departments as examples of other agencies.

The largest sources of funding for most fish and game departments include the sale of licenses and permits, followed by three major federal cost-sharing programs. All states require that hunters and fishermen purchase licenses before pursuing their sport. There are a host of different programs

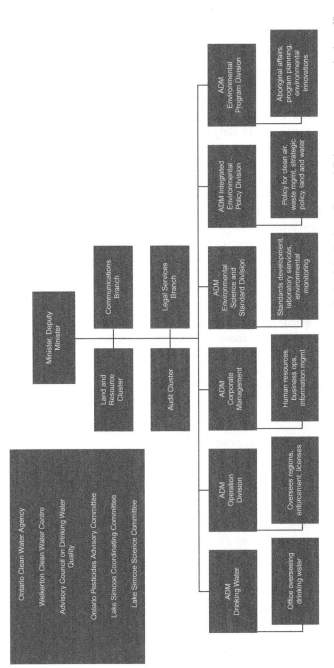

FIGURE 8.5 Partial organizational chart for the Ontario Ministry of the Environment. Each of the Assistant Deputy Ministers oversees several other offices, which do not fit on this chart.

among states to accomplish this. One method is to sell basic licenses that will allow a sportsman to harvest small game and common species of fish. In addition special permits may be required, such as habitat stamps, trout or walleye stamps, deer permits, migratory bird stamps, and the like. In the Midwest, deer hunting is enjoyed by many hunters and states often have split seasons with separate permits for each season or for each weapon (firearm, bow, muzzleloader) and even additional permits for does or for extra deer in some overpopulated counties. Coastal states may have some licenses for freshwater fishing and others for onshore fishing. In most cases big game licenses are sold at a premium rate, and resident sportsmen pay less than non-residents who want to hunt or fish in a particular state. For example, in a recent year a resident elk license cost around $570 and a non-resident license over $1000 in Wyoming. Across the states, stamps and licenses may account for up to 75% of an agency's income.[8]

The federal government has several programs to aid states in managing fish and wildlife. The two largest are the *Pittman-Robertson* or *Federal Aid in Wildlife Restoration Act* (1937), and the *Dingell-Johnson or Federal Aid in Sportfish Restoration Act* (1950). Both are 25% state, 75% federal cost-sharing programs. Half of the allocations are made on the basis of state land area, and half on population. Thus Wyoming, with a large area but relatively few people, and Rhode Island, with little land but twice the human population, can both obtain a fair share of these funds. States develop proposals that may include research, education or other activities for the funding. In 2011, the Wildlife Restoration Act brought in over $371 million and the Sport Fish Restoration over $349 million[9] to be apportioned over several programs. Over the past several years, the Wildlife Restoration funding has been spent on the following:

- Surveys and investigations (22%)
- Land acquisition (16%)
- Development (42%)
- Coordination (5%)
- Planning (1%)
- Technical assistance (1%)
- Hunter Education and Safety (13%).

The Sport Fish Program dollars are spent on the following programs:

- Sport Fish Restoration (57%)
- Aquatic Resources Education (15%)
- Boating Access (15%)
- Boating Infrastructure (3%)
- Clean Vessel (3%)
- Coastal Wetlands (6%)
- Coordination (1%).

Funding for both programs comes from excise taxes spent on fishing and hunting equipment. For every gun, box of ammunition, piece of reloading equipment, bow or archery equipment, 10% to 11% of the purchase price is returned as an excise tax specifically for Federal Aid in Wildlife Restoration.[9] Similarly, each fishing rod, tackle box, lure or other fishing equipment purchase returns 10% of the cost to the US Fish and Wildlife Service for disbursement to the states. Together, the programs have brought in more than $14 billion since their inception.[10]

Another more recent program is the *State Wildlife Grant*.[11] This program, established in 2003, is funded via direct allocation by Congress. At its inception, the program required that states complete a comprehensive conservation plan that identified the types of sites, species of concern and major conservation needs within each state. Its funding is based on 33% due to land area and 67% due to state population, and includes species other than game animals, and habitats, as potential funding needs. In 2011 the program provided over $47 million for wildlife and fisheries conservation, but during the years when states conducted their planning additional funds were allocated; thus the total amount distributed in the 10 years after 2002 exceeded $667 million. There are other federal programs that provide financial or technical support to states. Some of these are provided by the US Fish and Wildlife Service; others are provided by the US Forest Service (see Chapter 7).

Tax check off, lottery and similar programs can be very beneficial if they can be passed by the respective legislatures. Arkansas has a law that allocates one-eighth of 1% of income tax to conservation. While this seems a very small amount, over a 10-year period the tax brought in more than $415 million.[12] Way back in 1992, Colorado passed a law that allocates 50% of Colorado state lottery revenues among the Division of Parks and Outdoor Recreation, the Division of Wildlife, non-profit conservation organizations, and local governments. This has amounted to $2.4 billion since inception, and currently runs at around $59 million per year.[13] Fish and wildlife conservation benefits from one-eighth of 1% sales tax in Missouri, which provides about $85 million per year or 59% of the Department of Conservation's budget.[14] In 1998 the Virginia state legislature enacted House Bill 38, which provides that 2% of the state's sales tax collected on hunting, fishing, and equipment related to wildlife activity is appropriated to the Department of Game and Inland Fisheries; this amounts to around $12 million annually.[15] While these special allocations can be exceedingly helpful, relatively few states have enacted such laws, possibly because of the competing needs each state faces. Within the scope of special funds, several states have established conservation or wildlife automobile license plates (Figure 8.6) where an extra fee is charged that goes to the conservation agency. On average only about 10% of a state's fish and wildlife management costs are covered by general state funds, and many states do not allocate any general funds, expecting that the agency should be self-sufficient.

FIGURE 8.6　An example of a state conservation license plate. Part of the proceeds from the plate go to the state's conservation agencies.

FUNDING FOR NATURAL RESOURCES IN CANADIAN PROVINCES

Due to the Canadian concept that provinces take the lead role in natural resource management and enforcement, the federal government provides support for provincial programs through cost-share partnerships, where the federal and provincial governments each contribute to solve mutual problems. However, the bulk of funding for natural resource management is largely up to the provinces. Like the states, Canadian provinces charge fees for hunting and fishing privileges, some parks have user fees, and fines are charged to those who violate pollution or other environmental laws. Extraction fees and permits for mineral resources, gas and oil help to support natural resource management, sometimes by up to several million dollars. In both Canada and the United States, additional support for natural resource management may come from non-government organizations (see Chapter 9).

Study Questions

8.1 In what ways do the constitutional structures of states and provinces lead to similarities and differences in how they manage natural resources?

8.2 Describe some of the regulations that states and provinces can use to regulate wildlife and fishery harvests.

8.3 Discuss some of the general forms of natural resource administration at the state or provincial level. What are the advantages and disadvantages of these forms?

8.4 Discuss the roles of the Governor, Commission and Director in administering a specific agency.

8.5 Describe the role of Federal Aid to Wildlife Restoration, Federal Aid to Sport Fish Restoration, and the State Wildlife grant in funding state fish and wildlife management. Do Canadian provinces have similar programs that they can rely on? Why, or why not?

8.6 What other ways do states and provinces have of funding natural resource activities?

REFERENCES

1. USA Govt. State Government. <http://www.usa.gov/Agencies/State-and-Territoriess. html>; 2013.

2. State of Illinois. *Illinois Digest of Hunting and Trapping Regulations*. Springfield, IL: Illinois Department of Natural Resources; 2012.

3. Canada Privy Council.<http://www.pco-bcp.gc.ca>.

4. Neimanis VP. Crown Land. The Canadian Encyclopedia. <www.thecanadianencyclopedia. com>; 1999.

5. US Commission on Ocean Policy. <http://govinfo.library.unt.edu/oceancommission/>; 2005.

6. Lueck D. *An Economic Guide to State Wildlife Management*. Gardners, PA: PERC Research Study RS-002, Wildlife Management Institute; 1997.

7. Management Assistance Team. A Comparison of the Organizational Structure of State Fish and Wildlife Agency Commissions and Boards. Association of Fish and Wildlife Agencies. <http://matteam.org/reports/CB%20Report%20-Summary%20and%20Raw%20Data.pdf>; 2010.

8. Wildlife Management Institute. *Organization, authority and programs of state fish and wildlife agencies*. Washington, DC: WMI; 1987.

9. US Fish and Wildlife Service Wildlife and Sport Fish Restoration Program. <http://wsfrprograms.fws.gov/>.

10. Vocus Public Relations. <http://us.vocuspr.com/>.

11. US Fish and Wildlife Service. State Wildlife Grant. <http://wsfrprograms.fws.gov/Subpages/GrantPrograms/SWG/SWG.htm>.

12. Anonymous. Arkansas 1/8th-Cent Conservation Sales Tax 10-Year Report <http://www.agfc.com/aboutagfc/Documents/consvtax_tenyr_report.pdf>; 2007.

13. Colorado Lottery. Where the money goes. <http://www.coloradolottery.com/giving-back/where-the-money-goes/>; 2012.

14. Missouri Office of Administration. Budget & Planning. <http://oa.mo.gov/bp/budreqs2013/Conservation/Conservation.pdf>.

15. McMillen SL, Duda MD. *House Bill 38 and Future Directions for The Department of Game and Inland Fisheries: Results of constituent and staff studies and recommendations for future action*. Centreville, VA: Virginia Department of Game and Inland Fisheries; <http://www.dgif.virginia.gov/about/HB38_Final.pdf >; 2000

Part IV

Non-Governmental Agencies, People, and Money

Chapter 9

A Selected List of Non-Governmental Organizations (NGOs)

- Non-governmental organizations
- Not-for-profit
- Take free status
- Environmental Activism
- Monkeywrenching
- Eco-terrorism
- Ecotage

INTRODUCTION

Non-governmental organizations (NGOs) are just what the name indicates — organizations that are not formally part of government at any level. Many NGOs work closely with governments and use government grants to conduct their business, but they are independent associations. It would not be a stretch to say that there are NGOs for just about any issue or topic imaginable. NGOs range in size from a few million members to the proverbial local birding club. According to that font of wisdom *Wikipedia*, there are around 1.5 million NGOs in the United States.[1] Some of these NGOs are primarily lobbying groups that sponsor certain issues but do not qualify for tax-exempt status because a significant portion of their activities is associated with lobbying. In these cases, the primary NGO may have a foundation that is a tax-free organization to accept donations. For example, there are 5 million members of the National Rifle Association, which lobbies extensively for firearm owners' rights, but the affiliated NRA Foundation accepts tax-deductible donations.[2] Many other NGOs qualify for one of the several tax-exempt status categories in the tax code, such as 501 (c) (3). Still other NGOs are fronts

D.W. Sparling: Natural Resources Administration. DOI: http://dx.doi.org/10.1016/B978-0-12-404647-4.00009-X

for corporations or industries to provide a sense of "respectability" or a "cover".

With regard to natural resources, it may be impossible ever to know how many NGOs exist because there are multitudes that operate at very local levels and do not even apply for Facebook or Linkedin recognition (imagine that!). While some of these organizations officially register with the Internal Revenue Service or Canadian Revenue Agency, others are not significantly involved with funds and see no need to do so.

So what value do these NGOs have in natural resources? Again, it is difficult to be all-inclusive in this regard, but many of the activities of NGOs have real benefits and sometimes detriments, depending on the perspective.

1. At the most basic level, NGOs can be a place where people with some common interest can socialize.

2. They can serve as common areas where people with similar interests can associate – either actually or virtually – and provide some support for their common interest. Valid NGOs have memberships that comprise individuals, corporations or nations.

3. They can provide a unified, grassroots voice on specific issues. Legislators are more prone to listen to an organization that has 100,000 members than to single voters. Of course, if 100,000 voters independently wrote letters to their Congressman supporting an issue that would also be significant, but NGOs can and do organize their members and the votes!

4. NGOs can take action. While many spend considerable time trying to get governments to support their causes, others take an active participation in conservation by encouraging with voice, and with projects to recycle, restore habitat, conserve energy, rescue animals and other similar forms of hands-on participation.

5. NGOs can be where the average citizen cannot. There are numerous international organizations that conduct conservation projects in developing countries where the average citizen just cannot go. Members can vicariously participate in these activities through their contributions, and by reading the magazines or journals.

6. These organizations can facilitate government action. Government agencies are constrained by a host of regulations and laws, and often they cannot work expeditiously because of these constraints. NGOs do not operate under the same sets of rules and can act more quickly – for example, in purchasing an ecologically valuable piece of land.

7. Groups can and do serve as government watchdogs. Some of these organizations make sure that the various government agencies are doing what they have been established to do, while others may seem to be more interested in ascertaining that government is not overstepping its bounds.

8. NGOs serve as environmental watchdogs. Experts and members keep a focus on environmental issues and notify others when problems appear.

9. NGOs may counteract the activities of other NGOs. Keep in mind that NGOs reflect all attitudes and perspectives. Some actually work to diminish or rescind protections on the environment.

10. Sometimes NGOs can try to get public support by appearing to be something they are not. As we will see, some NGOs have environmentally friendly names to cover up actions that are otherwise.

In the pages that follow, we'll look at some of the not-for-profit NGOs that are associated with natural resources. While this list of organizations represents certain sectors of natural resources, it is far from being comprehensive. There are many more that could be added, and I am certain that some readers will wonder why one NGO is included and another is not. I have no real defense in this − your selection is probably as valid as mine. I hope, however, that these descriptions will enhance a sense of appreciation for the various types of NGOs and what they do.

For the purposes of trying to infuse some organization into this list, I've separated the NGOs into different categories. These categories include international organizations, individual species- or taxon- based groups, mega conservation groups, professional societies, and activist organizations. There can be some overlap in the interests of these groups. For example, some professional societies have connections to more than one country, and might be considered to be international. I have tried to distinguish these NGOs based on their primary stated interests. There is also brief mention of a couple of extremist groups and a few environmental imposters to fill out the slate.

A few words about the information in the boxes may help.

Chapters: some NGOs have local chapters at which members can meet and work at a local basis; others have regional offices but no local chapters *per se*; still others have both, and this category tries to be inclusive.

Funding: all the NGOs listed here are *not-for-profit*, meaning that they are corporations or associations that conduct operations for the general public without shareholders or a profit motive. Most also fall into one of the *tax-free* categories of the US tax code. Virtually all are open to charitable donations. You will see a lot of similarity in how they obtain funding.

Administrative/program ratio: there are three broad expense categories for not-for-profit NGOs − administration, fund raising and programs. Of these, the amount that is spent on programs is most important when considering an NGO. According to Charity Watch,[3] agencies should spend at least 60% of their income on programs. If an NGO has a smaller program allotment or a large administrative overhead, they might not be operating efficiently.

INTERNATIONAL ORGANIZATIONS

International NGOs may be headquartered in the United States, Canada or elsewhere, but have an international perspective on the conservation of natural resources. Some, such as IUCN, do not solicit individuals to become members but have nations and corporations as members instead.

International Union for Conservation of Nature

IUCN at a Glance

- **Organization:** International Union for Conservation of Nature and Natural Resources (IUCN)
- **Started:** 1948
- **Number of Members:** >1200 member organizations, including 200 + government and 900 + non-government organizations
- **Chapters:** 45 offices around the world
- **Funding:** Donations; funding from governments, bilateral and multilateral agencies, foundations, member organizations and corporations
- **Administrative/Program Ratio:** N/A
- **Focus:** Conservation of biodiversity especially endangered species around the world
- **Main Publications:** *IUCN Red List of Threatened Species*; many books and publications on global conservation
- **Headquartered:** Gland, Switzerland
- **Website:** www.iucn.org

The *International Union for the Conservation of Nature and Natural Resources* is a truly international organization with headquarters in Switzerland and offices on all of the continents. The IUCN is the oldest international conservation organization. It is most noted for its publication of the *Red List of Threatened Species*, which provides the status of thousands of species of plants and animals throughout the world. Its mission "is to influence, encourage and assist societies throughout the world to conserve the integrity and diversity of nature and to ensure that any use of natural resources is equitable and ecologically sustainable". The IUCN also focuses on promoting the value of nature through biodiversity conservation; effective and equitable governance of the use of nature through promoting people-to-nature relations; and deploying nature-based solutions to global challenges in climate, food and development to tackle problems of sustainable development.

World Wildlife Fund

WWF at a Glance

- **Organization:** World Wildlife Fund (WWF)
- **Started:** 1961
- **Number of Members:** 6.2 million globally
- **Chapters:** > 71 offices around the world
- **Funding Sources:** Donations, government grants, foundations, royalties
- **Administrative/Program Ratio:** 6.2%/73.0%
- **Recent Net Assets:** $271.7 million
- **Focus:** Conservation of wildlife and nature around the world
- **Main Publications:** Briefing papers, fact sheets, newsletters, booklets
- **Headquartered:** Washington, DC; Gland, Switzerland
- **Web:** https://worldwildlife.org

The *World Wildlife Fund* is based in Switzerland and Washington, DC with offices scattered across the globe including Canada. Its focus from the start has been on the conservation of rare and endangered animal species and their habitats. Like many internationally based natural resource organizations, however, the WWF has also become involved in enhancing ecological sustainability and promoting ecosystem values by working with businesses and industry. The WWF's mission is "to conserve nature and reduce the most pressing threats to the diversity of life on Earth",[3] and its vision is "to build a future where people live in harmony with nature".

The WWF works with local peoples and communities to promote a sense of responsibility and stewardship. "WWF's unique way of working combines global reach with a foundation in science, involves action at every level from local to global, and ensures the delivery of innovative solutions that meet the needs of both people and nature."[4]

Conservation International

Conservation International at a Glance

- **Organization:** Conservation International
- **Started:** 1987
- **Number of members:** Not a member-seeking organization
- **Chapters:** More than 30 offices internationally
- **Funding:** Donations, corporate sponsorships, foundations
- **Administrative/Program Ratio:** 11.1%/80.6%

- **Recent Net Assets:** $248.7 million
- **Focus:** Conserving ecosystem values and sustainability, often in developing nations
- **Main Publications:** Briefing papers, longer reports, focused issue books, no magazines
- **Headquartered:** Arlington, VA
- **Website:** www.conservation.org

The mission statement for *Conservation International* (CI) is "Building upon a strong foundation of science, partnership and field demonstration, CI empowers societies to responsibly and sustainably care for nature, our global biodiversity, for the well-being of humanity."[5] CI provides guidance to enable companies to reduce their impact on critical habitats and create economic and sustainable opportunities for local communities. Also, working with governments, civic organizations, and communities in developing countries allows CI to protect vast tracts of habitats while maintaining human rights and livelihoods. It works extensively in developing ecotourism sites, educating local populations on sustainable agriculture and other practices, and establishing demonstration sites so that others may witness conservation practices in work.

The Nature Conservancy

The Nature Conservancy at a Glance

- **Organization:** The Nature Conservancy
- **Started:** 1950; 1962 in Canada
- **Membership:** >1 million
- **Chapters:** Operations in all of the states, provinces and 35 countries around the world
- **Funding:** Donations, foundations, government grants, land sales, corporate support
- **Recent Net Assets:** $5.18 billion
- **Administrative/Program Ratio:** 12.4%/78.8%
- **Focus:** Conservation of land and water; a primary activity is acquiring ecologically valuable sites which can then be turned over to government agencies for permanent protection
- **Main Publications:** *Nature Conservancy Magazine*, state magazines or newsletters
- **Headquartered:** Arlington, VA
- **Website:** www.nature.org

The primary "business model" of *The Nature Conservancy* (TNC) is to find and preserve, through purchase, unique and ecologically important aquatic and terrestrial habitats in the world. To help donations and membership dues in funding more land purchases, TNC often sells acquired lands to organizations such as state or federal agencies or uses them in ecotourism and turns that money around. TNC can act more quickly than government agencies in acquiring land, and does a great service in preserving hundreds of thousands of acres. For TNC in the United States, the mission is "to conserve the lands and waters on which all life depends."[6] In Canada the statement reads: "The Nature Conservancy of Canada will lead, innovate and use creativity in the conservation of Canada's natural heritage. We will secure important natural areas through their purchase, donation or other mechanisms, and then manage these properties for the long term."[7] Over the past several years the NGO has expanded to protect lands against invasive species and human-induced perturbations, and conduct research to enhance understanding of these rare sites.

SPECIES- OR TAXON-FOCUSED GROUPS

Species- or taxonomic-based NGOs support particular species or groups of related species, such as waterfowl in the instance of Ducks Unlimited. While most of those listed here support hunting or fishing, there are many that focus on non-harvestable charismatic megafauna, such as the American Bald Eagles Foundation (www.baldeagles.org), the Cougar Network (www.cougarnet.org) or the International Wolf Center (www.wolf.org). There are also organizations dedicated to less predacious species, such as Hummingbird.net, the National Opossum Society (www.opossum.org), the promotion of frogs (Frogland, www.allaboutfrogs.org), or the North American Butterfly Association (NABA; www.naba.org). It is safe to say that this category has the widest representation of any.

Ducks Unlimited

Ducks Unlimited at a Glance

- **Organization:** Ducks Unlimited
- **Started:** 1937
- **Number of Members:** 596,300 +
- **Chapters:** Hundreds of local chapters, 84 university chapters
- **Note:** Ducks Unlimited Canada has 143,000 members
- **Funding:** Donations, membership dues, conservation easements, events, major endowments, federal and state habitat funds, royalties, sales
- **Recent Net Assets:** US$95.8 million; C$5.7 million

- **Administrative/Program Ratio:** 2.8%/77%
- **Focus:** Waterfowl conservation in North America including habitat restoration and acquisition, research, hunter education
- **Main Publications:** *Ducks Unlimited Magazine*
- **Headquartered:** Memphis, TN; Stonewall, Manitoba
- **Website:** www.ducks.orgwww.ducks.ca

Ducks Unlimited (DU) is the premier NGO for the waterfowl enthusiast. It has chapters all across North America, and has an unwavering focus on ducks and geese. Its mission statement is: "Ducks Unlimited conserves, restores, and manages wetlands and associated habitats for North America's waterfowl. These habitats also benefit other wildlife and people."[8] Its vision statement is similar: "The vision of Ducks Unlimited is wetlands sufficient to fill the skies with waterfowl today, tomorrow and forever." A majority of DU's members are hunters, and DU offers several services to them. There are annual meetings, hunter education courses, information on the biology of waterfowl, and a shop that offers a selection of hunting apparel.

Pheasants Forever

Pheasants Forever at a Glance

- **Organization:** Pheasants Forever
- **Started:** 1985
- **Number of Members:** 125,000 +
- **Chapters:** 600 +
- **Funding:** Endowments, investment interest, membership dues, easements, royalties, sales
- **Recent Net Assets:** $21.8 million
- **Administrative/Program Ratio:** 3%/88.8%
- **Focus:** Ring-necked pheasant and quail conservation (through sister organization Quail Forever), particularly habitat improvement, hunter and landowner education, and public awareness
- **Main Publications:** *Pheasants Forever Journal, Upland Tales, Quail Forever Journal*, newsletters
- **Headquartered:** St Paul, MN
- **Website:** www.pheasantsforever.org

Pheasants Forever and its newer sister organization, *Quail Forever*, focus on the conservation and increase of ring-necked pheasants (*Phasianus colchicus*) and northern bobwhite (*Colinus virginianus*). Although there are several species of quail in the United States, the organization is based in the Midwest and does not appear to have branched out to other species as yet. Both organizations are grassroots conservation groups whose mission is "Pheasants Forever is dedicated to the conservation of pheasants, quail and other wildlife through habitat improvements, public awareness, education and land management policies and programs."[9] Pheasants Forever claims a unique business model of being a strong chapter-based organization with headquarters empowering county and local chapters with the responsibility for determining how their locally raised conservation funds will be spent. Chapters provide assistance including training, manpower, and guidance to local landowners on habitat improvement.

Trout Unlimited

Trout Unlimited at a Glance

- **Organization:** Trout Unlimited
- **Started:** 1959
- **Number of members:** 140,000 +
- **Chapters:** 400
- **Funding:** Membership dues, events, grants, donations
- **Recent Net Assets:** $22.0 million
- **Administrative/Program Ratio:** 2.9%/89.6%
- **Focus:** Conservation and protection of cold water fishes and their watersheds; education; advocacy
- **Main Publications:** *Trout Magazine*
- **Headquartered:** Arlington, VA
- **Website:** www.tu.org

Trout Unlimited is primarily focused on coldwater species of fish – mostly trout and salmon. Its mission statement is brief: "To conserve, protect and restore North America's coldwater fisheries and their watersheds"[10]; its vision statement promises that "By the next generation, Trout Unlimited will ensure that robust populations of native and wild coldwater fish once again thrive within their North American range, so that our children can enjoy healthy fisheries in their home waters." To accomplish this, TU's chapters conduct habitat and stream improvement projects – providing structure, spawning areas and the like – and manage watersheds to assure that water inflows are clean. They also provide educational opportunities and encouragement to youth.

Rocky Mountain Elk Foundation

Rocky Mountain Elk Foundation at a Glance

- **Organization:** Rocky Mountain Elk Foundation
- **Started:** 1984
- **Number of Members:** 196,000 +
- **Chapters:** 500 +
- **Funding:** Membership dues, royalties, donations, conservation easements, land sales, grants
- **Recent Net Assets:** $42 million
- **Administrative/Program Ratio:** 10.5%/86.2%
- **Focus:** Conservation of elk in the United States, including western states and reintroduction into eastern states; hunter education
- **Main Publications:** *Bugle Magazine*
- **Headquartered:** Missoula, MT
- **Website:** www.rmef.org

The *Rocky Mountain Elk Foundation* (RMEF) has a very simple mission statement: "to ensure the future of elk, other wildlife, their habitat and our hunting heritage."[11] To accomplish this mission it states that the members focus on conserving, restoring and enhancing natural habitats; promoting the sound management of wild elk; restoring elk to their native ranges; and educating members and the public about habitat conservation and hunting heritage. It is a grassroots organization with an active chapter program that is clustered in the West but also occurs in the northern tier of states, Kentucky, and Tennessee, where local elk populations have been restored. According to a web search, there is a RMEF Canada, but it apparently did not have a website at the time of the search. The foundation provides hunter education and conservation advocacy, supports scientifically sound predator management, and conducts habitat restoration projects in elk habitat.

MEGA CONSERVATION GROUPS

In this category I include large NGOs with more than 1 million members and a base in the United States. Some of these are focused on multiple conservation issues, but the National Audubon Society has always had particular strength in avian conservation. These NGOs may also have offices in Canada, but, by and large, NGOs that support Canadian conservation are headquartered in the United States.

National Resources Defense Council

NRDC at a Glance

- **Organization:** Natural Resources Defense Council
- **Started:** 1970
- **Number of members:** 1.4 million
- **Chapters:** Five offices — four in the United States, one in Beijing
- **Funding:** Membership dues, government grants, foundations
- **Recent Net Assets:** $197.4 million
- **Administrative/Program Ratio:** 6.5%/78.3%
- **Focus:** NRDC is an international advocacy group staffed by attorneys and scientists to fight global climate change, save endangered species, and reduce pollution
- **Main Publications:** *OnEarth Magazine,* issue briefs, fact sheets, legislative analyses
- **Headquartered:** New York, NY
- **Website:** www.nrdc.org

The strength of the *Natural Resources Defense Council* (NRDC) is litigation and lobbying. While they solicit memberships, they do not have active chapters in the sense that the sportsmen's organizations above have. Their professional staff consists largely of attorneys, and scientists, and they do considerable work with environmental law. Their mission statement is rather lengthy, not very specific, and broad — something, in all good humor, you might expect an attorney to write:

The Natural Resources Defense Council's purpose is to safeguard the Earth: its people, its plants and animals and the natural systems on which all life depends. We work to restore the integrity of the elements that sustain life — air, land and water — and to defend endangered natural places. We seek to establish sustainability and good stewardship of the Earth as central ethical imperatives of human society. NRDC affirms the integral place of human beings in the environment. We strive to protect nature in ways that advance the long-term welfare of present and future generations. We work to foster the fundamental right of all people to have a voice in decisions that affect their environment. We seek to break down the pattern of disproportionate environmental burdens borne by people of color and others who face social or economic inequities. Ultimately, NRDC strives to help create a new way of life for humankind, one that can be sustained indefinitely without fouling or depleting the resources that support all life on Earth.[12]

The NRDC works with businesses, elected officials, and community groups on global warming, clean energy, oceans, endangered wildlife and their habitats; curbing pollution; ensuring safe and clean water; and fostering sustainable communities. Most of its activities occur in the United States.

Sierra Club

Sierra Club at a Glance

- **Organization:** Sierra Club
- **Started:** 1892
- **Number of Members:** 1.4 million +
- **Note:** Sierra Club Canada was founded in 1963 and has around 10,000 + members
- **Chapters:** 64 + around 85 student chapters; Sierra Club Canada has 5 chapters across the country
- **Funding:** Donations, membership dues, royalties, corporate partnerships
- **Recent Net Assets:** $82.6 million
- **Administrative/Program Ratio:** 2.0%/89.6%
- **Focus:** Grassroots conservation with a broad focus on wildlife, habitat, water, and renewable energy
- **Main Publications:** *Sierra Magazine*, Sierra Club books for adults and children
- **Headquartered:** San Francisco, CA; Ottawa, Ontario (Sierra Club Canada)
- **Website:** www.sierraclub.org; www.sierraclub.ca

The *Sierra Club* is the oldest environmental organization in the world. It was founded by the famed naturalist, John Muir. Its mission statement

To explore, enjoy, and protect the wild places of the earth; to practice and promote the responsible use of the earth's ecosystems and resources; to educate and enlist humanity to protect and restore the quality of the natural and human environment; and to use all lawful means to carry out these objectives.[13]

reflects the broad list of conservation issues on which the NGO focuses. Looking at its website, the impression is that if there is anything to do with the environment, the Sierra Club is concerned. Its primary actions include alerting its large membership about current issues, requesting them to write to their elected officials, and sometimes becoming involved in peaceful demonstrations. The parent organization works only in the United States, but there is a Sierra Club Canada. Unlike some of the other large organizations, the Sierra Club takes a more focused approach towards environmental issues and appears to be less focused on economic issues such as ecosystem values or sustainability.

National Wildlife Federation

National Wildlife Federation at a Glance

- **Organization:** National Wildlife Federation
- **Started:** 1936
- **Number of Members:** 4 million +
- **Chapters:** Affiliates in 48 states; campus organizations at 70 + colleges
- **Funding:** Royalties, membership dues, magazine sales, corporate sponsorship, grants
- **Recent Net Assets:** $40.6 million
- **Administrative/Operations Ratio:** 7.7%/79.8%
- **Focus:** Primary focus on wildlife, but includes education on sustainable conservation on many fronts – alternative energy, habitat conservation, endangered species, emphasis on public awareness especially with children
- **Main Publications:** *National Wildlife Magazine, Ranger Rick, Ranger Rick Junior*
- **Headquartered:** Reston, VA
- **Website:** www.nwf.org

The National Wildlife Federation (NWF) was founded by another leader in the early conservation struggle – Jay "Ding" Darling. The mission statement, "National Wildlife Federation – Inspiring Americans to protect wildlife for our children's future",[14] is more of a slogan than a statement of what the organization does, other than saying that the focus of NWF is wildlife and their habitats. Because the organization is large its members can be found assisting with many conservation issues, including habitat restoration, working against global climate change, reintroducing American bison (*Bison bison*) into areas from which they have become extirpated and so on. The NWF focuses heavily on the education of children. They have two very popular magazines for young children, and a third that appeals to adults, preteens and teens.

Canadian Wildlife Federation

Canadian Wildlife Federation at a Glance

- **Organization:** Canadian Wildlife Federation
- **Started:** 1962
- **Membership:** 300,000 +
- **Chapters:** N/A
- **Funding:** Donations, memberships, sales, grants, corporate partnerships

- **Administrative/Program Ratio:** 9%/67%
- **Recent Net Assets:** C$9.6 million
- **Focus:** Largest conservation organization in Canada; focuses on educating children and adults about wildlife conservation; advocacy, lobbying
- **Main Publications:** *Canadian Wildlife Magazine, WILD Magazine, Your Big Backyard Magazine,* newsletters, posters, guides
- **Headquartered:** Ottawa, Ontario
- **Web:** www.cwf-fcf.org/en/

The Canadian Wildlife Federation (CWF) is Canada's largest conservation organization. It is independent of the National Wildlife Federation in the United States, but the two organizations have mutual interests and have agreed to cooperate with each other. "The Canadian Wildlife Federation is dedicated to ensuring an appreciation of our natural world and a lasting legacy of healthy wildlife and habitat by: informing and educating Canadians; advocating responsible human actions; and representing wildlife on conservation issues."[15] The CWF has developed school curricula, a program to certify wildlife gardens, and educational materials including three well-illustrated magazines.

National Audubon Society

National Audubon Society at a Glance

- **Organization:** National Audubon Society
- **Started:** 1905
- **Number of Members:** N/A
- **Chapters:** ~ 500; offices in 24 states; 47 Audubon Nature Centers and sanctuaries
- **Funding:** Membership dues, sale of field guides, other royalties, endowments, bequests, investments
- **Recent Net Assets:** $400 million
- **Administrative/Program Ratio:** 6.9%/75.6%
- **Focus:** Conservation, primarily on birds but including habitat designations; clean energy; education and public awareness; conducts national Christmas and spring bird counts; operates sanctuaries and nature centers
- **Main Publications:** *Audubon Magazine,* Field Guide Series, *Audubon Adventures* curriculum guide
- **Headquartered:** New York, NY
- **Website:** www.audubon.org

The mission statement for the *National Audubon Society* (NAS) is "To conserve and restore natural ecosystems, focusing on birds, other wildlife, and their habitats for the benefit of humanity and the earth's biological diversity."[16] Like the Sierra Club and National Wildlife Foundation, the National Audubon Society was founded by yet a third conservation pioneer, George Bird Grinnell, and was named for John James Audubon, a famous painter of birds. From its inception the NAS has been a leading ornithological society for professionals and non-professionals. The society's activities include:

- engaging their local chapters nationwide in grassroots conservation action
- employing environmental policy, education, and science experts as advocates and instructors
- providing guidance to lawmakers and government agencies
- designating 2500 Important Bird Areas to identify, prioritize and protect vital bird habitat
- sponsorship of the Audubon's annual Christmas Bird Count, the new Coastal Bird Survey, and other initiatives to provide data on trends and status of birds
- supporting ecosystem-wide conservation initiatives to protect and restore some of the nation's special ecological features
- establishing Audubon Centers and sanctuaries for conservation, research and public awareness
- providing educational programs and materials and publishing *Audubon* to introduce schoolchildren, families and nature-lovers to nature and the power of conservation at home and around the world.

Each of its chapters is an independent organization, so obtaining an estimate of the total membership is difficult.

PROFESSIONAL SOCIETIES

Professional societies represent professionals such as wildlife biologists, foresters, fisheries biologists and others engaged in the conservation profession. Most sponsor regional, state or provincial chapters; hold annual meetings or conventions where professionals can share ideas and results of studies; and produce one or more journals that publish peer-reviewed research articles, commentary or position statements of conservation items relevant to the society. Many also have student chapters located at universities and are involved with undergraduate and graduate career development. If there is one primary shared purpose, it would be professional development. Membership dues are a main source of funding, but some also obtain corporate sponsorship, donations and other pockets of financing.

The Wildlife Society

TWS at a Glance

- **Organization:** The Wildlife Society
- **Started:** 1937
- **Number of Members:** nearly 11,000
- **Chapters:** 8 sections, 56 chapters, 131 student chapters, 26 working groups
- **Funding:** Membership dues, donations, publication subscriptions and conferences, grants and bequests
- **Recent Net Assets:** $1 million
- **Administrative/Program Ratio:** 29.5%/70.5%
- **Focus:** The chief organization for wildlife professionals in the United States and Canada; provides forums and annual meetings for its members, issues position statements and technical reviews on conservation issues, provides a professional certification program and educational opportunities, collaborates with partners to advocate for science-based wildlife policy
- **Main Publications:** *The Journal of Wildlife Management, The Wildlife Professional, Wildlife Society Bulletin*
- **Headquartered:** Bethesda, MD
- **Website:** www.wildlife.org

The Wildlife Society is the primary organization for professional wildlife biologists in the United States and Canada. It has state chapters in all of the states, student chapters at many universities that offer wildlife programs, and regional sections, including Canada. The organization states that "Our mission is to represent and serve the professional community of scientists, managers, educators, technicians, planners, and others who work actively to study, manage, and conserve wildlife and habitats worldwide."[17] The Wildlife Society uses its position and policy statements to encourage government leaders to enact laws that promote wildlife and habitat management.

Society of American Foresters

Society of American Foresters at a Glance

- **Organization:** Society of American Foresters
- **Started:** 1900
- **Number of Members:** 12,500
- **Chapters:** 32 state chapters, 17 divisions within states, 31 + student chapters
- **Funding:** Membership dues, donations, corporate sponsors, publication subscriptions and sales

- **Recent Net Assets:** Not available.
- **Administrative/Program Ratio:** Not available.
- **Focus:** Promoting the forestry profession through education, student programs, annual meetings, publications, certification programs
- **Main Publications:** *Journal of Forestry, Forestry Source, Forest Science, e-Forester, Northern Journal of Applied Forestry, Southern Journal of Applied Forestry, Western Journal of Applied Forestry*, several books on professional forestry
- **Headquartered:** Bethesda, MD
- **Website:** www.eforester.org

The Society of American Foresters was founded by Gifford Pinchot, the Father of Modern Forestry in North America. It is the chief professional society for foresters and forest managers. Its mission statement reads:

The Society of American Foresters *(SAF) is the national scientific and educational organization representing the forestry profession in the United States. Founded in 1900 by Gifford Pinchot, it is the largest professional society for foresters in the world. The mission of the Society of American Foresters is to advance the science, education, technology, and practice of forestry; to enhance the competency of its members; to establish professional excellence; and to use the knowledge, skills, and conservation ethic of the profession to ensure the continued health and use of forest ecosystems and the present and future availability of forest resources to benefit society. SAF members include natural resource professionals in public and private settings, researchers, CEOs, administrators, educators, and students.*[18]

Frankly, that does a pretty good job of summarizing the organization.

American Fisheries Society

AFS at a Glance

- **Organization:** American Fisheries Society
- **Started:** 1870
- **Number of Members:** 9000 +
- **Chapters:** Currently in the United States, Canada and Mexico; 4 regional divisions, 48 state and provincial chapters, student subunits within chapters
- **Funding:** Membership dues, donations, government support, subscriptions and sale of publications
- **Recent Net Assets:** $5.7 million
- **Administrative /Program Ratio:** 10.7%/89.3%
- **Focus:** Professional development, education, and conservation activities associated with fish and their habitats

- **Main Publications:** *Transactions of the American Fisheries Society; North American Journal of Fisheries Management; Journal of Aquatic Animal Health; North American Journal of Aquaculture; Fisheries; Marine and Coastal Fisheries: Dynamics, Management, and Ecosystem Science;* books; local chapter newsletters
- **Headquartered:** Bethesda, MD
- **Web:** www.fisheries.org

The American Fisheries Society is the oldest and largest organization catering to the professional fisheries biologist, whether the biologist works with fisheries management, aquaculture or health, or in a freshwater or marine environment. The mission of the *American Fisheries Society* is "to improve the conservation and sustainability of fishery resources and aquatic ecosystems by advancing fisheries and aquatic science and promoting the development of fisheries professionals."[19] The AFS has three principal goals: (1) to be a global leader providing information and technical resources for the sustainability and conservation of fisheries resources; (2) to facilitate life-long learning through world-class educational resources at all academic levels, and provide training for practicing professionals in all branches of fisheries and aquatic sciences; and (3) to serve its members and fisheries, aqua-culture, and aquatic science constituencies to fulfill the mission of the Society.

Canadian Society of Environmental Biologists

Canadian Society of Environmental Biologists at a Glance

- **Organization:** Canadian Society of Environmental Biologists
- **Started:** 1958
- **Number of Members:** 186
- **Chapters:** Eight, distributed provincially
- **Funding:** Membership dues, paid advertisements in the newsletters
- **Recent Net Assets:** N/A
- **Administrative/Program Ratio:** N/A
- **Focus:** Promotion of natural resources, professional development
- **Main Publications:** Quarterly newsletters
- **Headquartered:** N/A
- **Web:** www.cseb-scbe.org

The *Canadian Society of Environmental Biologists* (CSEB) is a small organization that, frankly, looks like it could use a boost in membership and funding. I have included it in this section for two reasons. First, it is one of the few professional conservation societies that is almost entirely based in Canada. While it has a few members in the United States,[20] more than 90% are in Canada. More importantly, however, CSEB exemplifies the many small NGOs scattered throughout North America that serve an important but largely unrecognized service. "The Canadian Society of Environmental Biologists is a non-profit registered society, whose primary focus is to further the conservation and prudent management of Canada's natural resources based on sound ecological principles."[21] The society's objectives are as valid as those of much larger organizations, and include:

- furthering the conservation and prudent management of Canada's natural resources so as to minimize adverse environmental effects
- ensuring high professional standards in education, research and management related to resources and the environment
- advancing education of the public and protecting the public interest on matters pertaining to the use of natural resources
- undertaking environmental research and education programs of benefit to the community
- assessing and evaluating administrative and legislative policies having ecological significance
- developing and promoting policies that seek to achieve a balance among resource management and utilization
- fostering liaisons among environmental biologists working towards a common goal within governmental, industrial and educational frameworks.

ENVIRONMENTAL ACTIVIST GROUPS

There are different forms of *activism*, and most of the above groups support some degree of it — including encouraging their members to write letters to Congressmen and Op-Ed pages of newspapers. Sometimes they may also advocate peaceful demonstrations. There is also the opposite extreme, represented by some of the groups mentioned towards the end of this chapter, which includes conducting illegal, destructive and sometimes injurious activities in the name of conservation; these latter groups seem to believe that the ends do justify the means. The groups that are represented in this particular section are or have been intermediate in their fervor. They encourage their members to write letters, but they are also apt to stage sit-ins, demonstrations and activities that push the legal limits. Examination of their histories will show that when they were young they acted much like youth and were less concerned about where the line between legality and illegality occurred than they are today. Today both of these NGOs have assumed greater respectability and tout large memberships, so their platforms must be attractive.

Friends of the Earth International

FOE at a Glance

- **Organization:** Friends of the Earth International
- **Started:** 1971
- **Number of Members:** 2 million +
- **Chapters:** Highly decentralized organization with national offices in 74 countries, including the US and Canada
- **Funding:** Membership dues, sales, interest, donations
- **Recent Financial Assets:** $1317 — holdings and assets are maintained with the individual national units
- **Focus:** Ecological sustainability and social justice around the world; fights against corporate damage of the Earth's resources; advisor to the UN; seeking alternatives to fossil fuels, genetically modified foods and habitat loss
- **Main Publications:** *Living Green Magazine*, newsletters, position papers
- **Headquartered:** Amsterdam, The Netherlands
- **Web:** www.foei.org

Friends of the Earth (FOE) differs from most of the organizations already described in being highly decentralized. While it has a headquarters in The Netherlands, its resources are dispersed across the national members. FOE international says:

We are the world's largest grassroots environmental network, uniting 74 national member groups and some 5000 local activist groups on every continent. With over 2 million members and supporters around the world, we campaign on today's most urgent environmental and social issues. We challenge the current model of economic and corporate globalization, and promote solutions that will help to create environmentally sustainable and socially just societies.[22]

FOE organizations across the world vary in their activities and focal points. In developing countries, there is often a strong emphasis on social justice and economic sustainability. In the United States, the emphasis is on getting corporations to act in a more conservation-minded way – to "think green". In Canada, FOE objectives include research, advocacy, education, and cooperation among various partners.

Greenpeace

Greenpeace at a Glance

- **Organization:** Greenpeace
- **Started:** 1971

- **Number of Members:** 2.9 million +
- **Chapters:** Offices in 40 countries around the world, including the US and Canada
- **Funding:** Donations, membership dues, will not take corporate money, national offices support the international headquarters, royalties, grants from Greenpeace Fund
- **Recent Financial Assets:** The international office does not hold any meaningful assets from one year to the next – virtually all assets are held by national offices
- **Focus:** One of the most visible conservation organizations for the past 40 + years – fighting against nuclear and fossil fuels, genetically modified foods, pollution, climate change; has been active in fighting whaling; helps protect ocean habitats; advocates sustainable agriculture
- **Main Publications:** *Greenpeace Magazine*, position papers, newsletters, single-issue publications
- **Headquartered:** Amsterdam, The Netherlands
- **Web:** www.greenpeace.org/international/en/, www.greenpeace.org/usa/en/, www.greenpeace.org/canada/en

Greenpeace International is similar to FOE in that the organization and its resources are dispersed among many national divisions. The headquarters provides central coordination but holds minimal financial or real assets. Greenpeace started in 1971 when "Marie Bohlen casually expressed the idea over coffee one morning [to boycott a nuclear test by the United States by sailing to Amchitka Island Alaska, the test site]. But the people around her – a lose [*sic*] alliance of Quakers, pacifists, ecologists, journalists and hippies – weren't known for shrugging off the really big ideas."[23] Through the 1970s and early 1980s Greenpeace was noted for its non-violent boycotts, sit-ins and demonstrations, and sending ships and inflatable Zodiacs to fight against nuclear testing and the whaling industry. One of their ships, the *Rainbow Warrior*, was even blown up by the French secret service in 1985 to prevent Greenpeace from interfering with a nuclear test. Today, its US mission is "Greenpeace is the leading independent campaigning organization that uses peaceful protest and creative communication to expose global environmental problems and to promote solutions that are essential to a green and peaceful future".[24] In Canada, the mission (abbreviated) is:

Greenpeace is an independent global campaigning organization that acts to change attitudes and behaviour, to protect and conserve the environment and to promote peace by:

1. *creating an energy revolution to address the number one threat facing our planet: climate change*
2. *protecting the world's ancient forests and the animals, plants and people that depend on them*

3. *defending our oceans by challenging wasteful and destructive fishing, and creating a global network of marine reserves*
4. *campaigning for sustainable agriculture by rejecting genetically engineered organisms, protecting biodiversity and encouraging socially responsible farming; and*
5. *creating a toxic free future with safer alternatives to hazardous chemicals in today's products and manufacturing.*[25]

THE DARK SIDE: ULTRA-RADICAL NGOS AND IMPOSTERS

In completing this chapter we would be remiss if we did not say at least a few words about ultra-radical NGOs – those that go beyond legal limits, convinced that the ends do justify the means, regardless of what happens in between. One of the most radical "pro-environment" NGOs is *Earth First!* Earth First! claims to be a "mission", not an organization, and it does not have memberships *per se* – yet it identifies 12 "chapters" and will gladly accept donations. Its website (www.earthfirst.org) reads like a manifesto. It states:

Earth First! was named in 1979 in response to a lethargic, compromising, and increasingly corporate environmental community. Earth First! takes a decidedly different tack towards environmental issues. We believe in using all the tools in the tool box, ranging from grassroots organizing and involvement in the legal process to civil disobedience and monkeywrenching.[26]

"Monkeywrenching" involves "tree sits", where members perch in trees that are marked for cutting, preventing loggers from working. The organization provided information in its newsletters on tree spiking – driving spikes into trees destined to be harvested so that the chain saws would buck, possibly doing serious harm to the logger. Although there is no evidence implicating Earth First! in actually carrying out this activity,[27] it is a form of *"ecotage"* (ecological sabotage) used by others.

The *Earth Liberation Front* (ELF) arose from the most radical elements of Earth First! in 1992. It makes no apologies about being an *eco-terrorist* group, proudly displaying a history of the movement on its website (earth-liberation-front.org). The rhetoric is full of hate and paranoia. Several of its members have been arrested for arson and other crimes: "Prison validates an ELF spokesperson's credentials; they've earned their stripes".[28] Although it has no formal leadership structure, the organization exists worldwide. "Monkeywrenching" in this group often involves destruction of property. Members have set fire to an SUV dealership, luxury homes, a fire ranger station, a ski resort, logging headquarters, a building at Michigan State University for its involvement with genetically modified organisms,[29] and a host of other structures.

We would also be negligent, I believe, if we did not reveal some of the great pretenders — those organizations (which crop up now and then until discovered by the press) that have or had ecologically friendly sounding names but are fronts for corporations that damage the environment. The *National Wetlands Coalition* sounds like a group of environmentally minded people in favor of wetlands. Actually, according to its mission statement:

The National Wetlands Coalition is ... diverse group of private and public sector entities that have joined together to advocate a balanced federal policy *[emphasis mine] for conserving and regulating the Nation's wetlands. Members of the Coalition own or manage wetlands and other "waters of the United States" that are subject to federal jurisdiction under Section 404 of the Clean Water Act. Coalition members include local governments, ports, water agencies, the development community, agriculture groups, electric utilities, oil and gas pipelines and producers, the mining industry, banks, environmental and engineering consulting firms, and Native American groups.*[30]

Watch out for phrases like "balanced federal policy" — they usually mean "none at all". The group has fought hard *against* wetland protection laws that prevent building, draining, drilling or farming.[31]

The *Global Climate Coalition* suggests an NGO that might be concerned about global climate change. It was disbanded in 2002, but before then was one of the most outspoken and confrontational industry groups in the United States battling *reductions* in greenhouse gas emissions.[32] It worked extensively with a network of industry trade associations, property rights groups and fringe elements that believed global warming is a plot to enslave the world under a United Nations-led "world government".

If you think of conservation, you might think "wise use of natural resources". However, the *Wise Use* movement is an industry-led, anti-environmentalist program founded in the late 1980s. At that time it dealt with fostering timber and mining issues in the western US. Subsequently, the movement inspired spin-off groups that look like grass-roots community organizations but are really sponsored by major corporations such as timber, mining and chemical companies. The groups proclaim that the well-documented hole in the ozone layer doesn't exist, that carcinogenic chemicals in the air and water will not harm anyone, and that clear-cutting is necessary for healthy trees. Every state has "Wise Use" groups that disseminate misinformation about the purpose and meaning of environmental laws. To Wise Users, environmentalists are pagans, eco-nazis and communists who must be fought with shouts and threats.[33]

As a final example of these "wolves in sheep's clothing" we'll mention *Citizens for the Environment*, which was established in 1990. How eco-sounding can you get? The group has no connection with the environmentally friendly *Citizens Campaign for the Environment* or other such valid groups. Apparently it also no longer exists, since it was exposed by the

Pittsburgh Post-Gazette as a front for industries that wanted to get rid of the Clean Water Act. The *Post* found that the "organization" did not even have members; it was strictly a shill organization to promote anti-environmental rhetoric.

The bottom line to both of these types of groups is that if you are going to support an NGO, be sure you know what the organization is all about. By the way, a disclaimer – including the organizations in this subheading does not imply that the author supports them in any way.

Study Questions

9.1 What are the major types of non-governmental organizations listed in this chapter? Can you think of any other categories that could be listed?

9.2 Where do most of the species-focused NGOs receive their funding from?

9.3 Do you belong to any NGOs? If so, look at their websites and determine their mission, primary sources of funding, types of memberships and activities. If they are a not-for-profit that accepts tax-deductible donations, they are supposed to publish annual financial statements and annual reports. Find those on their website – how much is spent on fundraising, administrative costs and programs?

9.4 Discuss the benefits and harm that various NGOs do to natural resources.

9.5 Which of the listed NGOs:
  a. Focus on wildlife?
  b. Are more focused on habitat or land?
  c. Have a strong commitment to sustainability or ecosystem values?
  d. Have student chapters?
  e. Are ones that you might like to get to know better?

REFERENCES

1. Wikipedia. Non-governmental organization. <http://en.wikipedia.org/wiki/Non-governmental_ organization>; 2013.
2. National Rifle Association. <http://www.nrafoundation.org/>.
3. Charity Watch. <www.charitywatch.org>.
4. World Wildlife Fund. About us. <http://worldwildlife.org/about>; 2013.
5. Conservation International. Government policy. <http://www.conservation.org/how/policy/ Pages/default.aspx/>; 2013.
6. International Union for Conservation of Nature and Natural Resources. About us. <http:// www.nature.org/about-us/vision-mission/>; 2013.
7. Nature Conservancy Canada. Mission and values. <http://www.natureconservancy.ca/en/ who-we-are/mission-values/>; 2013.
8. Ducks Unlimited. About us. <http://www.ducks.org/about-du/the-wetland-conservation- mission>; 2013.
9. Pheasants Forever. Pheasants Forever Mission/Model. <http://www.pheasantsforever.org/ page/1/mission.jsp>; 2013.

10. Trout Unlimited. About us. <http://www.tu.org/about-us>; 2013.
11. Rocky Mountain Elk Foundation <http://www.rmef.org/>.
12. Natural Resources Defense Council. About NRDC: Mission statement. <http://www.nrdc.org/about/mission.asp>; 2013.
13. Sierra Club. Sierra Club Policies. <http://www.sierraclub.org/policy/>; 2013.
14. National Wildlife Federation. Our mission. <http://www.nwf.org/Who-We-Are/Our-Mission.aspx>; 2013.
15. Canadian Wildlife Federation. About the Canadian Wildlife Federation. <http://cwf-fcf.org/en/about-cwf/>; 2013.
16. National Audubon Society. About us. <http://www.audubon.org/about-us>; 2013.
17. The Wildlife Society. Mission. <http://www.wildlife.org/who-we-are/mission>; 2013.
18. Society of American Foresters. <http://www.safnet.org/about/index.cfm>.
19. American Fisheries Society. Mission statement. <http://fisheries.org/mission-statement>; 2013.
20. Ryan PM. The Canadian Society of Environmental Biologists: An introduction for newer members and a review for the rest. *CSEB Bulletin*. 1994;51:9−11.
21. Canadian Society of Environmental Biologists. The CSEB Organization. <http://www.cseb-scbe.org/index.html>; 2013.
22. Friends of the Earth International. <http://www.foei.org/en/who-we-are>.
23. Greenpeace International. Amchitka: the founding voyage. <http://www.greenpeace.org/international/en/about/history/amchitka-hunter/>; 2007.
24. Greenpeace USA. <http://www.greenpeace.org/usa/en/about/>.
25. Greenpeace Canada. <http://www.greenpeace.org/canada/en/about/>.
26. Earth First! <http://www.earthfirst.org/about.htm>.
27. Wikipedia. Earth First! <http://en.wikipedia.org/>; 2013.
28. Earth Liberation Front. <http://www.earth-liberation-front.org>.
29. Wikipedia. Earth liberation front. <http://en.wikipedia.org/wiki/Earth_Liberation_Front>; 2013.
30. Integrity in Science. National Wetlands Coalition. <http://www.cspinet.org/integrity/nonprofits/national_wetlands_coalition.html>; 2003.
31. Environmental Working Group. Swamped With Cash: Sidebar 5: What is the National Wetlands Coalition? <http://www.ewg.org/sidebar-5-what-national-wetlands-coalition>; 1996.
32. Center for Media and Democracy. Global Climate Coalition. <http://www.sourcewatch.org/index.php/Global_Climate_Coalition>; 2012.
33. Center for Media and Democracy. Wise Use Movement. <http://www.sourcewatch.org/index.php/Wise_Use>; 2012.

Chapter 10

Stakeholders, Clients, and Cooperators: Who Are They?

Terms to Know

- Demographics of participation
- Iron triangle
- Consumptive Use of Wildlife
- Non-consumptive use
- State Wildlife Grant Program
- Amerindians
- First Nations

INTRODUCTION

In various parts of this book we have talked about the users of natural resources, referred to by agencies as *stakeholders*, cooperators, clients, customers, the public or other names. In Chapter 11 we will discuss how and why agencies and NGOs strive to develop good relationships and lines of communication with their stakeholders. Very briefly, government agencies work for, and because of, the public. If it is a federal agency, their public consists of the American or Canadian taxpayers. For state or provincial agencies, the public normally consists of the residents of that state or province. Also included are visitors to the various public parks, forests, refuges or other sites, such as foreigners, who do not support the site financially through taxes. Managers may have oversight for the fish, wildlife, trees or other resources in their charge, but they do so only because society, which consists of people, values the resources and their conservation. In addition to taxes, people spend billions of dollars each year to enjoy natural resources in a whole assortment of ways. Most of the problems a manager or agency encounters are also caused by people. It is an accepted truism that "[n]atural resource management is 90 percent managing the public and 10 percent managing the resource".[1]

D.W. Sparling: Natural Resources Administration. DOI: http://dx.doi.org/10.1016/B978-0-12-404647-4.00010-6

In this chapter we will examine several aspects of the broad stakeholder community. We will start with describing general characteristics of people who enjoy renewable natural resources. After that we will discuss how specific factors of the stakeholders, such as gender, ethnicity, age, income status, and education, affect their perspectives of the environment and conservation.

GENERAL DESCRIPTION OF OUTDOOR AND NATURAL RESOURCE USE

Both the United States and Canadian governments publish summaries on the use and economics of conservation-related activities at the national level. Both governments publish summaries of fisheries about every five years. The United States does the same for wildlife, national parks, and forest use. However, probably because of the separation of authority between federal and provincial governments in Canada, the federal government publishes less often and irregularly on these resources. States and provinces are similarly inconsistent in their publication of user data, if they publish any at all. However, the US Fish and Wildlife Service publishes a five-year summary on activities within each state.[2] Because the quality of data from the United States is usually more recent and complete than that from Canada, we will concentrate our presentation on the US. Attitudes towards conservation and environmental use are probably not all that different between the countries, so inferences can be made for Canada based on American data.

Natural Resource Use in the United States

Recently the responsibility for collecting the data for the FWS five-year summaries has been delegated to the US Census Bureau, but these reports are still published by the FWS. These reports include the number of participants aged 16 years or older; selected demographic characteristics of these participants; and the effort and the dollar amount they expend on pursuing the outdoor activities of fishing, hunting and wildlife observation. The surveys rely on responses from actual users, not purchase data on licenses, because many people purchase licenses without using them and wildlife observation activities generally do not require licenses.

Long-Term Demographics of Participation

The FWS portrays its data based on an index where the number of participants in 1955, the start of the survey, is adjusted to 100 (Figure 10.1). Between 1955 and 1975, the number of hunters in the United States steadily increased. Hunter participation then experienced a long, steady decline from 1975 to 2006.[3] This decrease was in spite of the US population increasing at an accelerating rate. During this period, the number of people participating

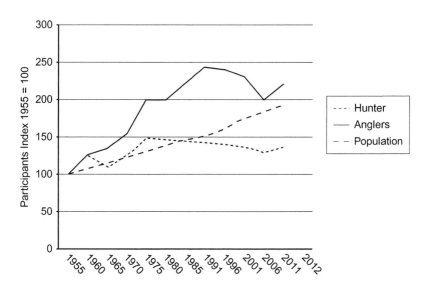

FIGURE 10.1   Participation by hunters and anglers in the United States from 1955 to 2011. The year 1955 is used as a benchmark and set at 100. Note the recent decline in both activities. *Data courtesy of US Fish and Wildlife Service.*

in fishing increased from 1955 to 1996 and then declined through 2006.[4] Declining fishing and hunting has caused considerable concern among the sporting public and agencies. Decreasing participation means fewer dollars for fish and wildlife conservation, because a substantial portion of funding to states comes from excise taxes on hunting and fishing equipment through the Fish and Wildlife Restoration Programs.[5] In addition, state fish and wildlife agencies receive a majority of their funding through the sale of state licenses and fees.[6] Moreover, hunting has long been recognized as a very important way of keeping some animal populations, such as deer and predators, in check. Uncontrolled ungulate populations can get out of hand and cause substantial harm to crops[7] and landscaping. Predator populations can reduce wildlife populations and inflict damage to livestock, so they may need thinning through sport hunting or trapping.[8] Hunting and trapping seasons, bag and season limits, and other harvest regulations are established to help manage these species.

While many agencies and professional wildlifers express concern about dwindling funding, some experts see the situation as an opportunity to enact change within more traditional state fish and wildlife departments.[5] For example, a common complaint about state agencies is that their management has been too focused on harvestable species. Critics call attention to an "*iron triangle*" of entrenched bureaucrats, policymakers and special interest groups that resist change and outside opinions[9] to keep things like they've always been. Euphemistically, these are called "the good old boys" who control the budgets of state fish and wildlife agencies. The critics argue that state and provincial agencies need to re-evaluate and set new goals, perhaps centered on the 95% of species that are non-game or "watchable wildlife", in order to succeed.

Several studies have attempted to determine the causes for the decline in hunting and fishing. Manfredo *et al.*[10,11] suggest that there has been a growing shift in attitude away from traditional wildlife values that emphasize use and management of animals for human benefits and towards one of animal rights. Animal rightists disagree with consumptive use of wildlife. This change in attitude is also demonstrated by a growing involvement in activities, such as wildlife observation, that do not involve harvesting fish or wildlife.[12,13] Other causes have been ascribed to an aging human population and declining birth rates (although, as can be seen in the Figure 10.1, the human population has continued to increase during this entire period); lack of time or finances; increased urbanization; perceptions dealing with the abundance of game, availability of hunting areas and over-stringent regulations; and a rise in virtual entertainment such as video games.[8,10,14–17]

Take another look at the right end of the graphs for the number of anglers and the number of hunters. You will see that there is a sharp tick upwards for both of these categories in the 2011 survey. This has caused some optimism in the conservation world.[18,19] The hope is that the long decline has slowed or even stopped. However, a single upward tick will need to be repeated more than once before we can claim a resurgence in outdoor activities.

In contrast to hunting and fishing, there does not seem to be any major decrease in non-consumptive enjoyment of nature. The number of visitors to National Parks over time, for example, shows a rather consistent increase over the years except for 1990 and 2006, when numbers dipped slightly (Figure 10.2). However, the rate of increase seems to be tapering off. Also, according to the FWS, the number of people involved in wildlife observation, whether from their homes or away, dipped about 15% between 1991 and 1996, but regained its levels of participation and essentially has remained stable since (Figure 10.3).

The National Forest Service has been collecting data on visitations to their sites only since 2005, and their system does not allow yearly comparisons. However, the annual number of visitations has varied only 0.5% since 2005, with a mean of 164.2 million visitors from 2005 to 2011.[20] It seems, then, that while the number of participants in consumptive use of wildlife and fish has declined through the years, non-consumptive use of natural areas has increased or remained stable in the United States.

Current Usage of Natural Areas in the United States

Annual participation in natural resource activities is a big deal in the United States. Each year millions of visitors use national and state parks, forests and

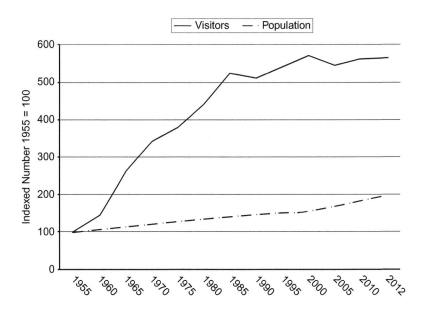

FIGURE 10.2   US National Park attendance 1955 to 2011 indexed for 1955 = 100. Unlike hunting and fishing, National Park attendance has enjoyed a steady increase – but the increase may be tapering off. *Data courtesy of US National Park Service.*

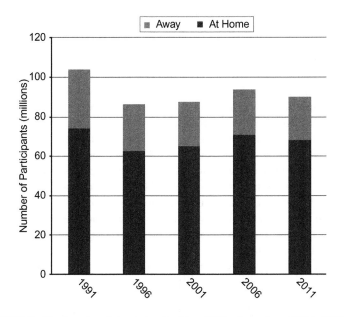

FIGURE 10.3   Number of participants involved in wildlife observation, 1991 to 2011. After a decline the numbers seem to have leveled out; however, that also means that the *per capita* visitation rate has declined. *Data courtesy of US Fish and Wildlife Service.*

Billions of Dollars

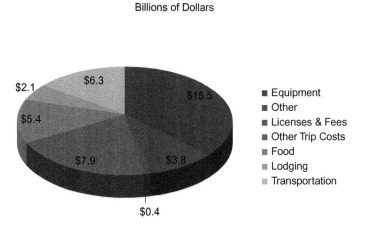

- Equipment
- Other
- Licenses & Fees
- Other Trip Costs
- Food
- Lodging
- Transportation

FIGURE 10.4   Expenditures in billions of dollars spent on fishing during 2011. The amount of money spent on equipment was by far the biggest cost. *Data courtesy of US Fish and Wildlife Service. This figure is reproduced in color in the color plate section.*

wildlife refuges for all manner of enjoyment, including hunting, fishing, wildlife observation, hiking, photography and canoeing. Participants spend literally billions of dollars while visiting these sites, with expenses including purchasing equipment, traveling, and costs directly associated with their sports. It is not possible to determine the precise number of individual visitors in any given year because many people visit more than one site. However, the total participation is staggering.

For example, the FWS estimates that 33.1 million people aged 16 years and older fished during 2011.[3] These people spent over $41 billion on their hobby (Figure 10.4). Of that, $15.5 billion was spent on equipment, which netted $1−1.5 million for the Federal Aid to Sports Fish Restoration program. Another $21.7 billion was spent on travel costs, helping local economies through lodging, fuel and food; and $400 million was spent on licenses and fees − most of which went to individual state agencies.

Nearly 14 million hunters spent $34 billion during the same year (Figure 10.5). Of this, $14 billion went to equipment, $0.8−1.2 million to the Federal Aid in Wildlife Restoration program, more than $10 billion in travel and to local economies, and $1 billion, through the more expensive licenses and fees, to states.

Non-consumptive observation of wildlife can be separated into those that enjoy wildlife on or near home and those who travel away from home, such as going on vacation to one of the National Parks. For this category, the FWS only includes instances where watching wildlife was the principal activity − not anglers or hunters who happened to watch wildlife while participating in their respective sports. In 2011, nearly 72 million people

Billions of Dollars

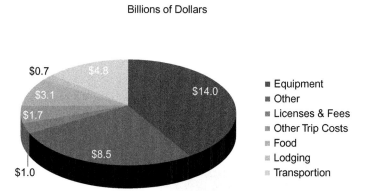

FIGURE 10.5   Expenditures in billions of dollars spent on hunting during 2011. Again, equipment was the biggest cost. *Data courtesy of US Fish and Wildlife Service. This figure is reproduced in color in the color plate section.*

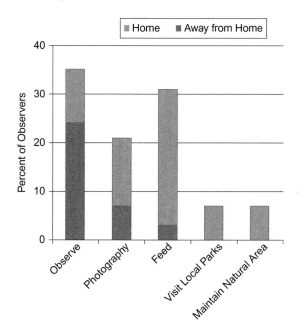

FIGURE 10.6   Number of people watching wildlife in the United States during 2011. The numbers are separated into those who watched near their home and those who traveled to do so; many more observed from home. *Data courtesy US Fish and Wildlife Service.*

watched wildlife in various ways (Figure 10.6) and spent $55 billion doing so (Figure 10.7). Observations around the home were three times greater (68 million vs 22 million) than those who traveled. Of course, many people probably did both – it is not really likely that somebody is going to travel

Billions of Dollars

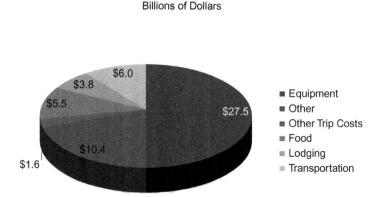

FIGURE 10.7 Total expenditures for people watching wildlife in the United States during 2011. As with hunting and fishing, the greatest cost was equipment. The *per capita* cost for watching wildlife is smaller than the other activities, but the much larger number of people who watch than fish or hunt makes up for the difference in total costs. *Data courtesy of the US Fish and Wildlife Service. This figure is reproduced in color in the color plate section.*

1000 miles to watch wildlife and not look out into their backyard once in a while – so the sum of home and away participants exceeds the total number of participants.

Comparatively, the number of people observing wildlife more than doubled those who fished and quintupled those who hunted. However, while the total amount of money spent by wildlife observers exceeded that for hunters or anglers, the amount spent per participant was substantially less for wildlife observers. Other than binoculars, guide books, bird seed and an occasional trip, wildlife observers don't have to spend a lot of money per person to enjoy their hobby, so the mean amount spent per observer averaged $765. In contrast, hunters spent a mean of $2484 and anglers $1261 per person. Although non-consumptive wildlife observers support local communities and conservation in many other ways,[21] very little if any of their expenses goes directly to wildlife or fisheries management. Wildlife-watching purchases do not entail an excise tax for Federal Aid programs, and observers do not have to buy licenses for their sport. This seeming inequality has caused criticism that wildlife watchers may not be paying their fair share for wildlife conservation. As a result, much of state wildlife conservation is paid for by hunters and anglers, thus maintaining a longstanding bias towards conservation of the 5% or so of all species that are harvestable.[22] Part of this difference was offset in 2002 by the State Wildlife Grant Program, which provides federal support for state management of species of concern, including non-harvestable species (see Chapter 12).

The overall impact of outdoor recreationists in the United States is immense. Combined, they spend over $130 billion a year on their hobbies.

With Federal Aid taxes, licenses, and fees, $4 billion goes directly to supporting management at state agencies, and billions are spent on other conservation measures.

Canadian Perspectives

As Canada has only about 11% of the population of the United States, we can expect that the total number of participants in outdoor activities and the amount of money that they spend are considerably less than that in the United States. While this may be so, Canadians as a people love the outdoors and are very active in visiting their public lands.

Natural Resource Participation in Canada

Because Canada has not kept records on natural resource usage for as long as the United States, long-term data are scarce; however, it appears that Canada is experiencing a similar drop in consumptive use of wildlife as the United States. For example, if we look at the number of recreational anglers in the country, data are available only from 1995 to 2010.[23] The Ministry of Fisheries and Oceans divides freshwater anglers into residents and visitors, because Canada draws a lot of anglers from the United States and other countries. Both sectors experienced a notable drop from 1995 to 2005, but the estimates in 2010 appear to be on par with 2005 (Figure 10.8).[18] Data on the number of hunters are primarily maintained by the provinces in Canada, and it is difficult to obtain comparable data across all provinces. However, there are data on migratory bird hunters from 1975[24] which show a 67% and a 72% decline in non-waterfowl and waterfowl hunting, respectively, over 35 years (Figure 10.9).[25] A similar situation is occurring among waterfowl hunters in the United States.[26]

Several reasons have been suggested for the decline, including tighter harvest restrictions, low waterfowl abundance (true in the past but not presently), and the cost and difficulty of waterfowl hunting compared to other forms of hunting; apparently all of these factors have contributed to diminishing hunting effort. Statistics indicate that hunters are turning to other species in lieu of migratory birds. For example, hunting statistics in Alberta show a general increase in several categories commensurate with the decline in waterfowl hunting.[27]

Another example of a decrease in participation at National Parks and other nature sites in Canada can be seen in Figure 10.10. There aren't many years in the record, but peak attendance occurred in 2008, dropped in 2009 and 2010, and remained stable in 2011.

Unfortunately, there isn't recent information on revenues produced by natural resources in Canada. The most recent data were collected in 1996 (Figure 10.11). At that time, C$12.7 billion came from several forms of

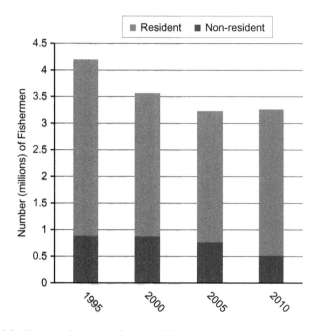

FIGURE 10.8 Number of anglers in Canada, 1995 to 2010, by resident and non-resident status. As with in the United States, angler participation in Canada declined over the 15 years, with possibly a slight upswing in 2010. *Data courtesy of Fisheries and Oceans Canada.*

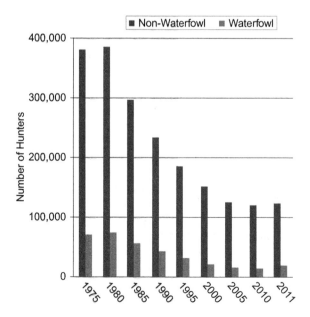

FIGURE 10.9 Number of migratory bird hunters in Canada, 1975 to 2011. There was an over 70% decline in the number of migratory bird hunters during that time. *Data courtesy of Statistics Canada.*

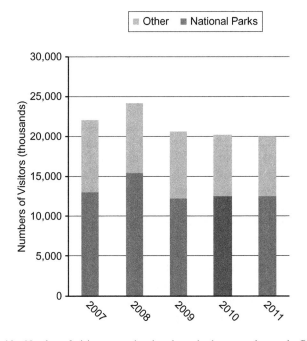

FIGURE 10.10   Number of visitors to national parks and other natural areas in Canada, 2007 to 2011. Although there was a brief increase in visitations during 2008, the overall pattern has been one of gradual decline. *Data courtesy of Parks Canada.*

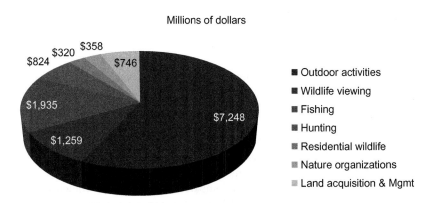

FIGURE 10.11   Total expenditures on fishing, hunting and wildlife observation in Canada during 1996. *Data courtesy of Statistics Canada. This figure is reproduced in color in the color plate section.*

participation. If we adjust for the 38% inflation rate in Canada over that time, the amount comes to C$17.5 billion today. Given the significantly smaller population in Canada compared to the United States, that works out to least an equivalent *per capita* outlay – perhaps a little more.

FACTORS THAT AFFECT PARTICIPATION OF STAKEHOLDERS

Now we have seen that millions of people participate in natural resource activities and that the money they spend is extremely important for maintaining conservation in both the United States and Canada, it may be of some value to examine intrinsic characteristics of stakeholders that affect their perspectives on conservation, expectations of management, and use of natural resources.

Some stakeholder communities can be very homogeneous. For example, a hunting club may consist of members of the same gender, ethnicity, age category and economic status. However, other stakeholder communities are diverse and are made up of individuals with different ideals and desires. It is always risky to pigeonhole individuals and draw conclusions about what a specific person's attitudes might be, based on the group anyone puts him/her into; some people just do not fit stereotypes, while others fall into multiple categories. As a result, surveys look for parsimony, trends or average responses, and often ignore the variance in the survey responses – it is just easier that way. Having said that, however, we are going to discuss several studies which have examined factors that potentially shape an individual's attitudes. In other words, does being a member of a particular gender, ethnicity, age group, income bracket or education level influence a person's attitudes towards natural resources? How can agencies reach children of different backgrounds and encourage them to take an active part in conservation?

Gender Differences

Gender differences towards natural resources have been extensively examined, and many interpretations have materialized. If you disagree with some of these interpretations, however, please do not shoot the messenger (me!).

Studies have shown that men are more likely to participate in natural resource activities than women. In 2011, 11% of all men and only 1% of all women hunted any wildlife species. The total number of hunters, or 13.7 million, was skewed 89%:11% men to women. Approximately 21% of all men aged 16 years or older fished compared to only 7% of women in 2011. Put another way, 73% of those who fished in that year were men and 27% were women, out of a sample population of 33.1 million.[28] However, the ratios were reversed for wildlife observation (Figure 10.12). In this case 27% of all men and 30% of all women watched wildlife, with women out numbering men 54%:46% of 68.6 million surveyed.

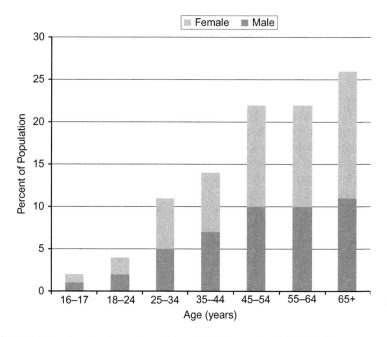

FIGURE 10.12   Percent of people watching wildlife by gender, 2011. While men do more hunting and fishing, women seem to be more active in observing wildlife. *Data courtesy of US Fish and Wildlife Service.*

In Canada during 2010, 3.3 million people aged 15 years and older fished[29]; 73% of the resident anglers were male and 27% were female. These numbers are consistent with studies which show that women are less likely to be consumptive users of fish and wildlife, and more likely to be non-consumptive viewers, than men.

The factors that affect participation in hunting are complex,[30] and often differ between the sexes. For one thing, most men get started in hunting as boys through their fathers or a male relative like an uncle, whereas women are more often initiated at a later age by their husbands.[31,32] For men, being outdoors and experiencing the thrill are primary reasons for hunting. For women, though, being with their husband in a sport he liked was a primary motivator to become involved with hunting. Because women are indoctrinated at a later age and their reason is tied more to a person than to the sport, women tend to be less committed to hunting then men[33] and less likely to spend large sums of money on their sport. Women raised in rural environments are much more likely to have at least tried hunting than those raised in urban areas, but men are even more influenced by growing up in rural environments.[34]

We cannot rely only on the sale of licenses to determine sex ratios or other characteristics of hunters. In Texas, although 111,000 licenses were sold to women in 1993, only 30% of those women actually responded to a harvest survey form.[35] Follow-up contacts suggested that 33% of the women receiving licenses admitted that they did not hunt – apparently they were obtaining extra deer permits for their family members. A similar situation occurred in Alberta[26] in 1995, and probably any situation where permits are issued without first requiring a hunting license runs the same risk of abuse.

As far as other gender differences towards resources are concerned, women are statistically less accepting of hunting then men.[36] Women have also been found to be less knowledgeable than men about wildlife and conservation issues, especially regarding threatened and endangered species, although they are equally knowledgeable about domestic species. Women tend to be more concerned about aesthetically attractive and evolutionarily more advanced animals than men are; and while men tend to be more utilitarian in their attitudes, women are motivated more by emotional attachments to specific animals.[32] Women also tend to be more fearful of animals and more willing to avoid most forms of wildlife–human conflicts than men are.[37]

Another difference between men and women may be associated with the desire to bring home food, but not all the studies agree. A stronger desire by women to bring meat back home was supported in a study of fishing in Minnesota.[38] According to this study, women were less likely to fish than men, but when they did go out women were statistically more interested in bringing fish home than were their gender counterparts. While men often released all the fish they caught and traded out larger fish for smaller for the sake of the sport, women were more likely to keep all the fish they caught up to the creel limit, and to keep larger fish over smaller ones. Men were also more accepting of laws that limited creel and size, and ethical issues of catch-and-release, than were women. However, when it comes to hunting, some data appear to be conflicting. On the one hand, Decker et al. (2001)[39] suggested that enjoying the outdoors was a bigger incentive for women to hunt than actually harvesting anything. Conversely, Duda (2001)[40] cited evidence that women were actually more likely than men to want game to bring home. Burger (2000)[41] found that women were less likely to consume wild fish or game than men, although there was no difference in rates of consumption of domestic fowl, beef, pork or commercially sold fish. This reluctance in women to eat wild fish may, however, be related to greater concerns about higher levels of contaminant concentrations, such as mercury, in women than in men.

For most of the history of the United States and Canada, women have pretty much been considered secondary participants in natural resources. While this may be clear in the active participation in natural resource recreation, it is even more evident in historic patterns in employment, management and administrative roles. Today, while all major natural resource departments in the United States and Canada have had at least one woman director, it has often taken over a hundred years for them to get a woman leader. In this regard, we should mention that Canada was ahead of the United States in electing the first woman national leader. Prime Minister A. Kim Campbell took office in June 1993, but left office in November of that same year. As of 2013, Canada has not had another female Prime Minister, and the United States has never had a woman President.

Among the major natural resource departments in the United States, the first woman director to serve was E. Mollie Beatie, for the FWS, from 1993 to 1996. Jamie Rappaport Clark followed her from 1997 to 2001. The Department of the Interior has had two women as Secretary of the Department: Gale Norton (2001–2006) and Sally Jewell (2013–). Similarly, the National Park Service has had two women directors. The Forestry Service has had one Chief and one Associate Chief. Even the Bureau of Land Management, long a bastion of male dominance, has had a woman director. From 2010 until at least 2013 the acting director of the US Geological Survey has been a woman, as was her predecessor.

Environment Canada has had one woman minister, Leona Aglukkaq, who started office as the Minister of Environment Canada in 2013. Natural Resources Canada has had two women as ministers, and Parks Canada has had one woman.

There are thousands of women working in these various agencies in all sorts of positions – scientists, upper and middle management, biologists, foresters, technicians, administrative staff. However, the women mentioned here represent the sum total of top female leaders in federal natural resource agencies in both countries since their inception. It is clear that the gender ratio in the very top federal leadership positions is highly skewed. While that might be understandable over the early history of the agencies, due to archaic gender roles, the gender ratio in modern times remains skewed in favor of Anglo-American males.

The role of women in top leadership roles at the state or provincial level is not much better than at the federal level. I scanned the websites of the state agencies for departments of natural resources, fish and game, parks, and forestry, and found that only about half of the states had at least one woman in a senior leadership position as director, deputy director or commissioner of the agency. About 10 women were actually directors or chairpersons of a commission. At the same time, I found only one province that had a woman director of their Ministry of the Environment. The representation of women leaders in the private work place is comparable, and has been nearly constant at 13% to slightly over 14% in executive officer jobs of Fortune 500 companies since 2009[42] The bottom line, I suppose, is that the old glass ceiling separating women from top leadership positions in North American natural resources is beginning to bend, but it hasn't been shattered yet.

The disparity between men and women in leadership roles does not appear to be due to a lack of interest by women in environmental issues.

Women are at least as likely to be involved in formal and informal environmental organizations as men, and often lead local environmental movements against toxic and hazardous wastes, and environmental disasters such as Love Canal and Three Mile Island.[43] The traditional explanation for the disparity in women leadership, at least according to many feminists, is that the traditional roles of housekeeper and mother prevent women from being involved in leadership roles.[44] This might be a factor, but other studies have suggested that these are trade-offs that a lot of women are willing to make.[45] As a male scientist and not a sociologist, I am not going to make a fool of myself and comment too extensively on the topic. Rather, I'm going to take the prudent way and suggest that there are many reasons behind the skewed gender ratio in natural resource leadership positions. Personally, I don't see any reason not to include women as full and equal members in the leadership hierarchy.

Perhaps this quotation from the US Forest Service's document "History – Women in the Forest Service"[46] may be encouraging:

In recent years women have become involved in every aspect of national forest management, as well as employed in increasing numbers in the research, state and private forestry, and international programs. The most noticeable changes came in the dangerous duties of smokejumping, fire fighting, and law enforcement where women were finally admitted in the late 1970s and early 1980s. The facts are that women employed in the Forest Service are in every capacity and experience from the new recruits to the top administrative positions. It is hoped that soon the employment of women in any capacity in the agency will no longer be heralded as another first. They will be considered as the most qualified person for the job, no matter what the work entails.

Differences in Participation and Attitudes Due to Ethnicity and Race

In the 2011 survey conducted for the FWS, there were considerable differences in outdoor recreation across ethnicity and race (Table 10.1).[24] Less than 0.5% of Hispanics and less than 0.1% of African Americans and Asian Americans hunted. Fishing was more common, with 6% of Hispanics and 6% to 8% of the racial minorities wetting a line now and then. The greatest representation of minorities was those who fished only. By comparison, 17% of Anglo-Americans fished or hunted, 4% hunted and more than 11% fished. The other obvious aspect was that those who were classified as "White" formed a larger percentage of each participant group than indicated by their percentage in the total population. Hunting in the United States seems to be a "White" sport, while fishing has a greater representation among minorities but is still dominated by those classified as "White".

TABLE 10.1 Characteristics of Anglers and Hunters aged 16 years and over, by Ethnicity and Race.

Race or ethnicity	Total population (million)	% of population	% who hunted or fished	% of hunters or fishermen	% who fished only	% of fishermen	% who hunted only	% of hunters only	% who fished and hunted	% of fishers and hunters
Total population	239	100	16	100	10	100	2	100	4	100
Hispanic	32	14	6	5	5	6	< 0.5	3	< 0.5	2
Non-Hispanic	207	86	17	95	11	94	2	97	4	98
White	183	76	18	87	11	84	2	97	5	93
African American	23	10	10	6	8	8	< 0.1	< 0.1	2	4
Asian American	12	5	6	2	6	3	< 0.1	< 0.1	< 0.5	< 0.5
All others	21	9	8	4	6	5	< 0.5	2	1	3

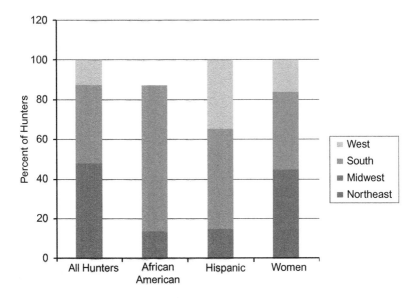

FIGURE 10.13 The distribution of minority anglers is consistent with the resident population, with the greatest number in the southern portion of the country. *Data courtesy of US Fish and Wildlife Service.*

Not unexpectedly, the regional distribution of minority hunters is approximately proportional to their distribution across the United States (Figure 10.13). As of 2010, 73% of African American hunters lived in the South and 14% in the Midwest; the other 13% were scattered in the Northeast and West.[47] Hispanics were more widely distributed across the South, Midwest and West. Excepting the low numbers of minority hunters, their patterns generally follow the hunting population at large. A majority of Hispanics and African Americans tend to hunt on private lands − 68% and 84%, respectively, but so do 82% in the general population. Most hunters prefer to hunt big game species, but African Americans are much more likely to pursue small game as well, and Hispanics prefer migratory birds compared to the general public. In 2010, African Americans spent a total of $198 million on hunting trips and equipment while Hispanics spent $354 million, compared to a total hunting expenditure of around $9.7 billion.

Of the $20 billion spent by anglers in that same year, African Americans and Hispanics spent $579 million and $817 million, respectively. Mean expenditures tend to increase with age, education, urbanization and income for all groups, but there is a really sharp increase in spending among Hispanics with a college degree compared to other levels of education, or other groups with college degrees. By far, most fishing occurred on freshwater. The amount of fishing by minorities tends to be greatest in the 16- to 24-year-old and 55+ years brackets. It decreases with education, is higher in the $20,000−50,000 income brackets compared to < $20,000 and > $50,000 brackets, and is considerably higher in rural areas than in urban locations. Most fishing occurs in the South, regardless of minority or general population. We will discuss some of these general trends in greater detail below.

Unfortunately, the publications in Canada's series on *The Importance of Nature to Canadians: The Economic Significance of Nature-related Activities*, which was published up to 1996, does not examine natural resource participation by ethnic, economic or age category, so there do not seem to be comparable statistics for Canada.

It is critical for agencies to be aware of minority opinions and use (or lack of use) of natural resources now and in the future. Currently at 10% of the total US population, Hispanics are rapidly increasing in numbers. Between 2000 and 2010, more than half of the population growth in the nation was among Hispanics; in 2010 they totaled 31.8 million, according to the US Census Bureau.[48] Hispanics are not an entirely homogeneous group − while Mexicans account for about 59% of the total, Latinos also come from Cuba, Central and South America, Spain and the Caribbean, and move into the conterminous US from Puerto Rico, a US territory. People from each national origin have unique stories, experiences and desires. Three-quarters of the Hispanic population live in the South and West; over 50% live in California, Texas and Florida. The areas with the most rapidly growing populations are the Midwest and South. Only 8.5% of Hispanics live in rural areas,[49] compared to 20% of Anglo-Americans. People in rural areas have more opportunities to hunt and fish than do those in urban areas.

Hunt and Ditton reported on predictions based on population dynamics that Hispanics may account for 16% of all anglers in the United States by 2025, and as much as 29% of freshwater anglers in Texas by the same date,[50,51] emphasizing again the importance of understanding the needs of this group.

Hunt and Ditton[52] also summarized information on Hispanic societal attitudes and made predictions on their approach to fishing based on that information. Once again this information was summarized from sample populations, consists of generalizations, and does not capture the range of attitudes held by Latinos. Some of their observations included the following.

1. Apparently, Anglo-Americans tend to have an attitude that they are dominant over nature, that humans are separate from nature, and that we can and should use nature for the "greatest good" of humans. Hispanics, in contrast, tend to view nature more fatalistically − what will happen will happen, and humans have little to no control over nature. Humans and nature are more united under this perspective. The authors interpreted this

by suggesting that fishing for Anglo-Americans may be more driven towards catching the "big one" and less oriented towards bringing home food. Compared to Hispanics, many Anglo-Americans appear to accept catch-and-release programs.

2. Anglo-American culture tends to be future-oriented, and revolves around schedules, promptness and timeliness so that future goals can be achieved. Hispanic cultures seem to be more present-oriented and less strongly tied to timeliness. The difference leads to a stronger urge among Anglo-Americans to separate work and leisure and to plan for future leisure time. Hispanics may be more opportunistic in their fishing adventures. Anglo-Americans are more driven by the excitement and feeling of achievement than are Latinos.

3. Hispanics appear to be more family-oriented than many Anglo-Americans. For them, fishing is an opportunity to take the family out and enjoy companionship. In contrast, Anglo-Americans characteristically focus on short-term goals with specific desired outcomes. Their emphasis is on activity and efficiency. Thus, fishing for them may be more of an expression of independent success, although there are elements of getting together with family and friends.

African Americans have been the largest minority group in the United States for many decades. In 2010, those that considered themselves Black, African American, or African American and some other category in the US Census numbered 42 million, or nearly 14% of the total US population.[53] From 2000 to 2010 the Black population grew at a faster rate than the Anglo-American population, so they, like Hispanics, are a group to watch through future years. The US Census Bureau recognizes 12 groups of Blacks, based on various combinations of Black, White, Hispanic or Other; whether there are any systematic differences among these groups in their perception of natural resources is dubious, but there are differences in regional distribution. Approximately only 8% to 10% of all African Americans live in rural areas.[54]

In a follow-up study, Hunt and Ditton[55] continued to profile Hispanics and Anglo-Americans and included African American anglers. They suggested that the absence of outdoor recreational opportunities in urban areas is a major factor leading to decreased involvement in such activities. This seems logical, for, as we have explained, a much larger percentage of Hispanic and African American populations live in urbanized areas compared to Anglo-Americans. Overall, the average annual income of Latinos and Blacks is less than that of Anglo-Americans, and the incidence of fishing increases with income to a point and then levels off. Fewer Blacks or Hispanics probably have access to power boats that facilitate fishing than do Anglo-Americans. Dwyer and Hutchinson[56] supported this supposition in a

Chicago-based study which showed that Blacks were significantly less likely to participate in boating activities than Anglo-Americans.

Human likes and dislikes, and consequently behavioral patterns, develop at early ages, and if children are not given opportunities to experience outdoor recreation they are less likely to pursue these experiences when they mature. Lack of interest in the outdoors, therefore, can be a self-perpetuating characteristic in minority cultures. Both Hispanic and African American cultures tend to be more group- or people-oriented than is Anglo-American culture,[57] and this again would make fishing more a family excursion than a solitary affair. Lower incomes might also encourage minorities to seek easily caught fish with larger creel limits for food than large game fish for trophies.

Hunt and Ditton[55] conducted a survey of Texas anglers to identify ethnic differences. The actual survey took place more than 20 years ago but cultural differences are slow to change and, because the general trends in minority fishing patterns have generally remained the same since Hunt and Ditton's study, as determined by the five-year analyses conducted by the FWS, the findings still have some applicability. The overall proportion of Texans purchasing fishing licenses in 1990 was 14.7%. However, among male and female Anglo-Americans the licensed percent was around 40%. For the rest of the population, licenses were purchased by 20% of the Hispanics and 14% of the Blacks. Surveyed Anglo-Americans tended to start fishing around 12 years of age and had about 28 years of experience, whereas Hispanics and African Americans started in their early 20s and had 3 to 5 years' less experience. All three groups fished about the same number of days in a year (mean range 21 to 40), but Anglo-Americans, significantly, spent about half of their days aboard a boat whereas minorities fished from banks about 75% of the time. Black bass (*Micropterus* spp.) were the preferred species, followed by crappies (*Pomoxis* spp.) and catfish. Men of all three groups stated that they fished more often with family and friends than alone. Consistent with hunting and fishing, far fewer minorities claimed to observe wildlife, either at home or away (Figure 10.14).

In Canada, the politically correct term for people of different color, not including aborigines, is "visible minorities". In contrast to the United States, the principal minority groups in Canada are from Asia, making up 68% of all visible minorities. Fifteen percent of the total Canadian population is visible minorities. I could find no information on their perceptions or use of natural resources in Canada.

In both the United States and Canada there are minorities of particular importance. These are the Amerindians, people of the First Nation, or aborigines. In the United States, 5.2 million people or 1.7% of the population claim Indian status. The US government recognizes 565 tribal governments, each of which can declare its own natural resource laws on land that it

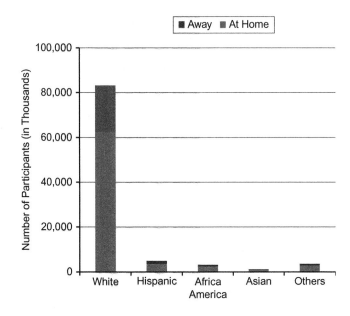

FIGURE 10.14   Like the proportion of minorities that hunt or fish in comparison to Anglo-Americans, the proportion that observes wildlife is also small. *Data courtesy of US Fish and Wildlife Service.*

possesses. Life on tribal lands — either independent nations or reservations — is difficult, with little employment opportunity, and only about 25% of the Amerindian people live on these lands.[58] However, traditional Amerindian culture is treasured in these areas, and various federal laws acknowledge Amerindian pre-eminence in dealing with use of wildlife and plants for subsistence, culture and religion. Outside of the tribal lands, Amerindians are generally treated as US citizens and are subject to US and state laws.

In Canada there are three groups of aboriginal peoples: First Nations, Métis, and Inuit. They have a combined population total of 1.4 million, about 4.3% of the Canadian population. Inuits primarily live in Nunavut, Northwest Territory, Northern Quebec (Nunavik) and Labrador. Many First Nations people live on reserves. First Nations, Métis and Inuit who live off of reserves account for 56% of all Canadian aborigines.[59] As in the United States, aborigines living on reserves have independence to establish their own laws concerning natural resources. Off-land, they are labeled non-status aborigines and are governed like everyone else.

Discerning minority leadership in natural resource agencies at the state or federal level is difficult because surnames are often misleading, photographs are not always available, and someone can be of mixed heritage and either acknowledge that or not. We do know that Manuel Lujan Jr (1989–1993), appointed by President George Herbert Walker Bush, was the first Latino secretary of the US Department of the Interior, for his ethnicity was well publicized. In 2013, Leona Aglukkaq was the first Inuk or Inuit to be appointed Minister of the Environment Canada. Prior to that she was the Minister of Health and, of course, a Member of Parliament. The majority of employees, including many of the top officials in the US Bureau of Indian Affairs of the Department of the Interior, are either native Americans or Eskimos.[60]

Stakeholder Age

Young people are the future. The attitudes and perceptions that youth form will affect the rest of their life. If they are convinced that conservation is essential, our natural resources have a chance for adequate management. If, on the other hand, they do not develop a respect for nature, the future can be very different. Unfortunately, the 5-year survey on hunting and fishing conducted by the FWS and Census Bureau does not include people less than 16 years of age. At least one study[61] indicates that there is a growing resistance against teaching children how to hunt. In 2005, the sports channel ESPN asked the question: How early should children be taught how to hunt? This was not a scientific poll, for responses were collected from 25,000 people who voluntarily called in. However, at least 40% of respondents in six states stated that children should never be taught how to hunt.

When actual data on hunting or fishing are absent, data on knowledge and attitudes can help to discern if youth at least appreciate nature, and might help predict whether they will be concerned about the environment when they become adults. One survey conducted on 464 fifth-graders in Minnesota[62] showed that they were generally informed about wildlife and conservation issues, but when asked if they practiced conservation at home the scores were a lot lower. A longitudinal 29-year study on nearly 100,000 high school students[63] showed that ups and downs in concern for the environment related to current events in each year. Some consistent patterns during that time included a decline in personal conservation-oriented activities. Students became significantly less concerned with curtailing driving and conserving energy through time. In 2005 students were less likely to believe that resources were in short supply compared to 1976. Although attitudes regarding the danger and amount of pollution varied through time, students demonstrated a moderate to high concern about pollution throughout the survey. High school seniors also demonstrated a general increase in materialism through the years. The authors suggested that this rise in materialism is a potential explanation for the general decline in conservation behavior. The demand for a higher standard of living including more material possessions is negatively related to behaving in a conservation manner, and is consistent with a declining belief that resources are limited but that technology will prevail. The increase in materialism may also be a threat to enjoying environmental aesthetics, for how can one compare a forest glade with the latest fashions?

There was one finding that is particularly relevant to government agencies. Students tended to be more willing to allow government to deal with environmental issues than to take personal responsibility. However, when environmental crises were perceived, both a reliance on government to take care of things and a sense of personal responsibility increased. This finding is important, for it indicates that high school students have trust in the government and may be willing personally to contribute to environmental health if provided with guidance by state and federal agencies.

For children and teens, virtual entertainment is increasingly becoming a major distraction from all forms of outdoor activity – in fact, for activity in general. Sociologists are coming to understand that it is not television that is the main distraction, but the use of video games, computers, pods, pads, smart phones, and the totality of electronic diversions that is most influential. In a well-cited study, Vandewater et al.[64] used data from two national surveys on 2831 children aged from 1 to 12 years. They determined that while television watching per se was not correlated with obesity in children, video game use was. Pergams and Zaradic[65] compared several predictor variables to a per capita decline in National Park visitations from 1988 to 2002. They found that this decline in per capita visitors was significantly and negatively correlated with several electronic entertainment indicators, including hours of television watching, video games playing, home movie and theatre attendance, and Internet use. Similarly, use of the Internet and video games were both negatively related to the amount of time spent hunting among adults. While video games are often associated with children, only 32% of those who play video games are aged under 18 years, and the mean age of video players is 30 years old.[66] However, it is just not the playing but the amount of playtime that is important. At least in one study, male children aged 10–15 years played a mean of 43 hours per week – equivalent to a full-time job – while females played 30 hours[67]; either way, a very substantial portion of the week can be spent on indoor, sedentary "activity".

Among adults surveyed by the 2011 National Survey of Fishing, Hunting and Wildlife-Associated Recreation, the proportion of people who either fish or hunt was smallest in the 16-year-old group, increased through the 45- to 55-year-old age group, and then decreased (Figure 10.15). Because the incidence of fishing has remained relatively stable over recent years, the pattern we seen in a given year can be at least partially explained by life situations at the time. Among 16- to 17-year-olds the proportion of hunters and anglers is small due to competing sports such football, soccer, or basketball, not to mention studies. Moving beyond high school, college activities and lack of spendable money can reduce the opportunity to participate. Post-college or in their 30s to 50s many people have more time for leisure pursuits, but after

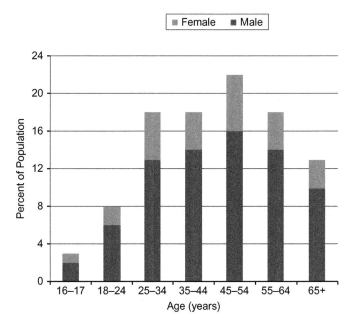

FIGURE 10.15   The proportion of hunters and anglers by age group in 2011. The general pattern is increased participation through the middle years, followed by a decline as the population ages. *Data courtesy of US Fish and Wildlife Service.*

that health issues may begin to reduce activity. The pattern in the graph is also partly determined by the decline in hunting over the past several years. There simply isn't a replacement population coming into the older age segment, so we also see attrition in the older age brackets.

Other Stakeholder Characteristics

The list of factors that influence stakeholder attitudes and behavior is seemingly limitless. However, to close this chapter there are two other factors that we should at least mention; income, and level of education of the stakeholder.

Regarding *income level*, we have already briefly mentioned the importance of income to outdoor activities. The FWS survey reported a bimodal curve in the relationship between outdoor sports and income level (Figure 10.16). Factors behind this curve are complex and interacting. For example, the < $20,000 income level seems to be somewhat of an anomaly, but the other low income levels include the elderly and the urban poor, who either have difficulty in getting around or lack the opportunity to fish and hunt. As income levels increase, people become more financially mobile and can travel to natural areas. The decrease in activity at the highest income

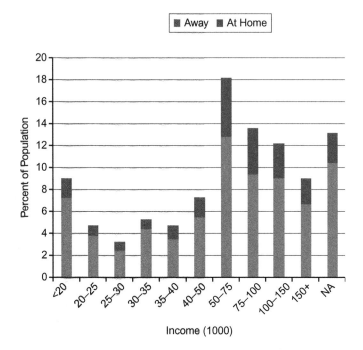

Income (1000)

FIGURE 10.16 Outdoor wildlife observation by gender and income level, 2011. Except for the group with the lowest income appearing to participate in wildlife, there is a general trend towards increased participation with increased income, followed by a dip in the highest income brackets. *Data courtesy of US Fish and Wildlife Service.*

levels may be due to the availability of other sources of leisure and entertainment, including power boating, travel, or similar activities. Beyond a doubt, income interacts with the other factors discussed, such as gender, ethnicity and age.

Regarding *level of education*, NSFHAR demonstrated that the peak of participation in hunting and fishing occurs among those who graduated high school and had some college (Figure 10.17). Those without a high school education could be either the youngest segment of the survey or the poorest. Again, they are likely to be urban poor with little opportunity to participate in outdoor activities. Beyond a couple of years at college, the proportion of those who hunt or fish declines with level of education. Interactions with income are likely in these categories, and they may have a wider set of leisure opportunities than others.

In sum, the average hunter in the United States is going to be a white, non-Hispanic 35- to 65-year-old male making around $50,000–100,000 per year, with a high school education and living in a rural area anywhere except along the coasts. The average angler will be similar, except that he will make between $40,000 and $75,000, could have an education anywhere from

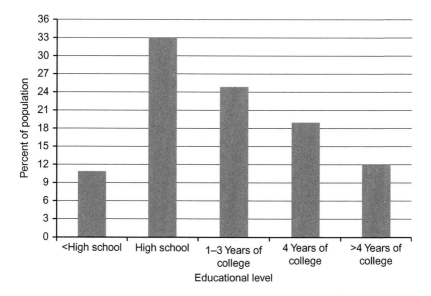

FIGURE 10.17   Distribution of people who hunt or fish in the United States by education level, 2011. Those with a high school diploma form the largest share of the pool. *Data courtesy of US Fish and Wildlife Service.*

high school to three years of college, and live in an urban environment just about anywhere in the United States. For the real sportsman who both hunts and fishes, look for someone who is white, non-Hispanic, 35 to 65 years old, makes $40,000−100,000, and has an education from high school through college degree. He probably lives in a rural environment somewhere in Wisconsin, Illinois, Indiana, Michigan or Ohio, but he could also live elsewhere as long as it is not the coastal or mountain states.

Study Questions

10.1   What factors were listed in this chapter as possibly affecting stakeholder perceptions and needs with regards to natural resources? Can you think of other factors that might influence stakeholders?

10.2   Do you agree with the list of differences between men and women as they pertain to the use and appreciation of natural resources?

10.3   Do you agree or disagree with the list of differences among ethnic and race groups as they influence their perceptions and involvement in natural resources?

10.4   Why do you suppose that there has been a dearth of women and minorities in top leadership positions? What should be done about it?

10.5   What reasons do you think are leading to an overall decline in consumptive use of fish and wildlife? Do you think that will pose long-term problems in conservation?

REFERENCES

1. Fazio JR, Gilbert DL. *Public Relations and Communications for Natural Resource Managers*. 2nd edition. Dubuque, IA: Kendall/Hunt Publishing Co; 1986.
2. US Fish and Wildlife Service. *2011 National Survey of Fishing, Hunting, and Wildlife-Associated Recreation*. State Overview. Arlington, VA: US Fish and Wildlife Service; 2012.
3. Aiken R. *Trends in Fishing and Hunting 1991–2006: A Focus on Fishing and Hunting by Species: Addendum to the 2006 National Survey of Fishing, Hunting, and Wildlife-Associated Recreation Report 2006-8*. Washington, DC: US Fish and Wildlife Survey; 2010.
4. US Fish and Wildlife Service. *2011 National Survey of Fishing, Hunting, and Wildlife-Associated Recreation: National Overview*. Washington, DC: US Fish and Wildlife Service; 2012.
5. US Fish and Wildlife Service Wildlife and Sport Fish Restoration Program. <http://wsfrprograms.fws.gov/home.html>.
6. Jacobson CA, Decker DJ. Ensuring the future of state wildlife management: understanding challenges for institutional change. *Wildl Soc Bull*. 2006;34:531–536.
7. Brown TL, Decker DJ, Riley SJ, et al. The future of hunting as a mechanism to control white-tailed deer. *Wildl Soc Bull*. 2000;28:797–807.
8. Way JG, Bruskotter JT. Additional considerations for gray wolf management after the removal of Endangered Species Act protections. *J Wildl Manage*. 2012;76:457–461.
9. Clark JR. Managing dissonance in the iron triangle. *The Freemen: Ideas on Liberty*. 1996;46:693–697.
10. Manfredo MJ, Pierce CL, Fulton D, et al. Public acceptance of wildlife trapping in Colorado. *Wildl Soc Bull*. 1999;27:499–508.
11. Manfredo MJ, Teel TL, Bright AD. Why are public values toward wildlife changing? *Human Dimen Wildl*. 2003;8:287–306.
12. Adams CE, Leifester JA, Herron JSC. Understanding wildlife constituents: birders and waterfowl hunters. *Wildl Soc Bull*. 1997;25:653–660.
13. Miller CA, Vaske JJ. Individual and situational influences on declining hunter effort in Illinois. *Human Dimen Wildl*. 2003;8:263–276.
14. Mehmood S, Zhang D, Armstrong J. Factors associated with declining hunting license sales in Alabama. *Human Dimen Wildl*. 2003;8:243–262.
15. Martínez-Espiñera R. Public attitudes toward lethal coyote control. *Human Dimen Wildl*. 2006;11:89–100.
16. Dempson JB, Robertson N, Cochrane M, et al. Changes in angler participation and demographics: analysis of a 17-year license stub return system for Atlantic salmon. *Fish Manage Ecol*. 2012;19:333–343.
17. Robison KK, Ridenour D. Whither the love of hunting? Explaining the decline of a major form of rural recreation as a consequence of the rise of virtual entertainment and urbanism. *Human Dimen Wildl*. 2012;17:418–436.
18. US Department of Interior Press Release. Salazar: Survey Delivers "Great News for America's Economy and Conservation Heritage"; 2012.
19. Theodore Roosevelt Conservation Partnership. National Survey Shows Rise in Hunting, Angling Trends. <http://www.trcp.org/media/news-article/national-survey-shows-rise-in-hunting-angling-trends#sthash.k1ndpohW.dpuf>; 2012.
20. US Forest Service. National Visitor Use Monitoring Results. USDA Forest Service National Summary Report Data collected FY 2007 through FY 2011. http://www.fs.fed.us/recreation/programs/nvum/nvum_national_summary_fy2011.pdf>; 2012.

21. Gambardello JA. *Wildlife-watchers spending big in Cape May, region.* Philadelphia, PA: The Inquirer; 2002.

22. Anderson LE, Loomis DK. Balancing stakeholders with an unbalanced budget: how continued inequalities in wildlife funding maintains old management styles. *Human Dimen Wildl.* 2006;11:455−458.

23. Fisheries and Oceans Canada. Survey of recreational fishing in Canada, 2010. Ottawa Canada. <http://www.dfo-mpo.gc.ca/stats/rec/canada-rec-eng.htm>; 2012.

24. Environment Canada General Harvest Statistics. <http://www.ec.gc.ca/reom-mbs>.

25. McCombie B.Waterfowler Numbers Take a Dive. American Hunter. <http://www.americanhunter.org/article.php?id=19606&cat=4&sub=8>; 2012.

26. Vrtiska MP, Gammonley JH, Naylor LW, et al. As waterfowl hunters decline so do funds for wetland conservation. The Wildlife Society News, Summer 2013. <http://news.wildlife.org/twp/2013-summer/as-waterfowl-hunters-decline/>; 2013.

27. My Wild Alberta.com. Annual sales statistics. <http://mywildalberta.com/BuyLicences/AnnualSalesStatistics.aspx>.

28. US Department of the Interior, US Fish and Wildlife Service, and US Department of Commerce, US Census Bureau. *National Survey of Fishing, Hunting, and Wildlife-Associated Recreation.* Washington DC: US Census Bureau; 2011.

29. Fisheries and Oceans Canada. *Survey of recreational fishing in Canada.* Ottawa: Fisheries and Oceans Canada; 2011.

30. Heberlein TA, Serup B, Ericsson G. Female hunting participation in North America and Europe. *Human Dimen Wildl.* 2008;13:443−458.

31. Jackson RM. The characteristics and formative expressions of female deer hunters. *Women in Nat Resour.* 1988;9:17−21.

32. Jackson RM, McCarty SL, Rusch D. Developing wildlife education strategies for women. *Trans 54th NA Wildl Nat Resour Conf.* 1989;54:445−454.

33. McFarlane BL, Watson DL, Boxall PC. Women hunters in Alberta: Girl power or guys in disguise? *Human Dimen Wildl.* 2003;8:165−180.

34. Heberlein TA, Thompson EJ. Socio-economic influences on declining hunter numbers in the United States 1977-1990. In: Csany S, Enrhaft J, eds. *Trans XXII Congr Internat Union Game Biol − The game and the man.* 1996:373−377. Sofia.

35. Adams CE, Steen SJ. Texas females who hunt. *Wildl Soc Bull.* 1997;25:796−802.

36. Kellert SR, Berry JK. Attitudes, knowledge, and behaviors towards wildlife are affected by gender. *Wildl Soc Bull.* 1987;15:363−371.

37. Gore ML, Kahler JS. Gendered risk perceptions associated with human−wildlife conflict: Implications for participatory conservation. *PLoS ONE.* 2012;7:e32901.

38. Schroeder SA, Fulton DC, Currie L, et al. He said, she said: Gender and angling specialization, motivations, ethics and behaviors. *Human Dimen Wildl.* 2006;11:301−305.

39. Decker DJ, Brown TL, Seimer WF, eds. *Human Dimensions of Wildlife Management in North America.* Bethseda, MD: Wildlife Society; 2001.

40. Duda MD. The hunting mind. Women and hunting. <http://www.responsivemanagement.com/download/reports/NAHWomen.pdf>; 2001.

41. Burger J. Gender differences in meal patterns: role of self-caught fish and wild game in meat and fish diets. *Environ Res.* 2000;83:140−149.

42. Catalyst. Statistical Overview of Women in the Workplace. <http://www.catalyst.org/knowledge/statistical-overview-women-workplace>; 2013.

43. Tindall DB, Davies S, Mauboulés C. Activism and conservation behavior in an environmental movement: The contradictory effects of gender. *Soc Nat Resour.* 2003;16:909−932.

44. Chafetz JS. The gender division of labor and the reproduction of female disadvantage: Toward an integrated theory. *J Fam Issues*. 1988;9:108–131.

45. Reed M. Marginality and gender at work in forestry communities of British Columbia, Canada. *J Rural Studies*. 2003;19:373–389.

46. US Forest Service. USDA Forest Service History – Women in the Forest Service. <www.fs.fed.us/aboutus/history/womens.html>.

47. Henderson E. *Participation and Expenditure Patterns of African-American, Hispanic, and Female Hunters and Anglers: Addendum to the 2001 National Survey of Fishing, Hunting, and Wildlife-Associated Recreation*. US Fish and Wildlife Service, Arlington, VA; 2004.

48. Ennis SR, Rios-Vargas M, Albert NG. The Hispanic population 2010. US Census Bureau <http://www.census.gov/prod/cen2010/briefs/c2010br-04.pdf>; 2011.

49. Effland ABW, Kassel K. Hispanics in rural America: The influences of immigration and language on economic well being. USDA Econ Res Serv 87-99; 1996.

50. Waddington DG. *Participation and expenditure patterns of Black, Hispanic and women anglers. Report 9103. Addendum to 1991 National Survey of Fishing, Hunting and Wildlife-Associated Recreation*. Washington, DC: US Department of the Interior; 1995.

51. Murdock SH, Loomis DK, Ditton RB, et al. Demographic change in the United States in the 1990s and the twenty-first century: Implications for fisheries management. *Fisheries*. 1992;17:6–13.

52. Hunt KM, Ditton RB. Perceived benefits of recreational fishing to Hispanic-American and Anglo Anglers. *Human Dimen Wildl*. 2001;6:153–172.

53. Rastogi S, Johnson TD, Hoeffel EM, et al. The Black population 2010. US Census Bureau <http://www.census.gov/prod/cen2010/briefs/c2010br-06.pdf>; 2011.

54. Johnson K. *Rural Demographic Change in the New Century: Slower Growth, Increased Diversity*. Durham, NH: Carsey Institute <http://www.carseyinstitute.unh.edu/publications/IB-Johnson-Rural-Demographic-Trends.pdf>; 2012.

55. Hunt KM, Ditton RB. Freshwater fishing participation patterns in racial and ethnic groups in Texas. *North Am J Fish Manage*. 2002;22:52–65.

56. Dwyer JF, Hutchinson R. Outdoor recreation participation and preferences by Black and White Chicago households. In: Vining J, ed. *Social Science and Natural Resource Management*. Boulder CO: Westview Press; 1990:49–68.

57. Simcox DE. Cultural foundations for leisure preferences, behavior, and environmental orientation. In: Ewert AW, Chavez D, Magill AW, eds. *Culture, Conflict and Communications in the Wildland–Urban Interface*. Boulder, CO: Westview Press; 1993:267–280.

58. US Census Bureau News Room. 2010 Census Shows Nearly Half of American Indians and Alaska Natives Report Multiple Races. <http://www.census.gov/newsroom/releases/archives/2010_census/cb12-cn06.html>; 2012.

59. Aboriginal Affairs and Northern Development Canada. Urban aboriginal people. <http://www.aadnc-aandc.gc.ca/eng/1100100014265/1369225120949>; 2013.

60. US Department of the Interior Indian Affairs. Who we Are. <http://www.bia.gov/WhoWeAre/BIA/>; 2013.

61. Tanger SM, Laband DN. Attitudes about children learning to hunt: Implications for game management. *Human Dimen Wildl*. 2008;13:298.

62. Tanner D. Fifth grader's knowledge, attitudes and behavior toward habitat loss and landscape fragmentation. *Human Dimen Wildl*. 2010;15:418–432.

63. Wray-Lake L, Flanagan CA, Osgood DW. Examining trends in adolescent environmental attitudes, beliefs, and behaviors across three decades. *Environ Behav*. 2009;42:61–84.

64. Vandewater EA, Shim M, Caplovitz AG. Linking obesity and activity level with children's television and video game use. *J Adolescence*. 2004;27:71–85.

65. Pergams ORW, Zaradic PA. Is love of nature in the US becoming love of electronic media? *J Environ Manage*. 2006;80:387–393.

66. Entertainment Software Association. Essential facts about the computer and video game industry. <http://www.theesa.com/facts/pdfs/ESA_EF_2013.pdf>; 2013.

67. Homer BD, Hayward EO, Frye J, et al. Gender and player characteristics in video game play of preadolescents. *Comp Human Behav*. 2012;28:1782–1789.

The Need for a Good Public Relations Department

INTRODUCTION

One of the cardinal realities for all employees of agencies and non-governmental organizations is that you don't work alone. Everyone has someone that they have to answer to — i.e., a supervisor of some sort — and many are people who supervise others. Even the Prime Minister or the President is responsible to the voters, Congress and the Judiciary. This means that individuals cannot just act alone and decide to do whatever they want if they work for a government agency. Suppose, for example, that you are a refuge manager or forest biologist and you have an issue, illustrated by the star-like image in Figure 11.1. The issue could be anything that you might encounter during your work — let's say that it's a desire to increase the number of overwintering geese on your site. You decide that a good way of doing this would be to provide 15 acres of standing corn for winter food. If you owned your own land, that probably would not be a problem — you would cultivate the ground, plant the seeds and allow the corn to grow with no one to tell you otherwise. As a government land manager, however, you could very easily find yourself constrained from acting freely. First, you are responsible to the facility where you work. Is increasing the number of

D.W. Sparling: Natural Resources Administration. DOI: http://dx.doi.org/10.1016/B978-0-12-404647-4.00011-8

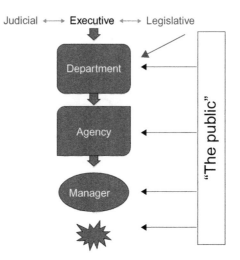

FIGURE 11.1 Land and wildlife managers have a series of authorities that affect their work and whether they can accomplish something that they want to do. These authorities develop policy, provide budgets, and ultimately determine the overriding objective and legality of the agency. At all levels, the public has input.

waterfowl part of the long-term strategic plan (see Chapter 12)? Next, you are responsible to the agency and the policy established by the agency. If you are working on a refuge for the Canadian Wildlife Service you might not have any problems in establishing your plot of corn, for it could very likely be within the policy of the agency. However, if you are in a National Park under the US National Park Service you will find it very difficult to persuade your supervisors, because planting a cornfield for waterfowl is unlikely to be within the agency's policy. In turn, the agency is responsible to the Department, whether it is Environment Canada or the US Department of the Interior, and both establish additional policy. They, in turn, are part of the Executive Branches of their respective governments, and are affected by the Legislative Branch for funding and direction through laws and by the Judicial Branch that determines if what the agency is doing is legal.

Note, however, the big box to the right in the diagram, labeled "*The Public*". Since the 1970s, this box has increasingly become more important. Prior to that time the public generally was not very interested in what went on in natural resource agencies. As long as they had a place to hunt, fish or recreate, most people were content. Since then, through a series of federal laws and a general increase in interest by the Canadian and American people, public opinion has become very important and interacts with government conservation policy and activities at all levels. For instance, it is quite possible that the public was the origin of the issue at the bottom of the figure, either by a desire for enhanced recreation or by causing a problem in the first place. Natural resource managers encounter the public on an almost daily basis, and the rapport that they develop can make a big difference in their ability to effect change. The public also interacts with agencies and departments in many ways. First, if the agency is supported by taxpayer or donor dollars, the public is in essence the "employer" of all of the people in the agency. Second, laws such as the National Environmental Policy Act and the Canadian Environmental Assessment Act (see Chapter 4) require public input into any major activity that involves the environment. Third, people today are more willing to express their opinions to agencies and to their legislators. Overall, it is not much of an exaggeration to say that "Natural resource management is 90 percent managing the public and 10 percent managing the resource".[1] Thus, a very important but sometimes overlooked component of any natural resource agency should be a *public relations department* that is experienced in working with the public and promoting the agency's agenda.

Public relations (PR) is "the business of inducing the public to have understanding for and goodwill toward a person, firm, or institution".[2] It can also be defined as "the planned effort to influence public opinion through good character and responsible performance, based upon mutually satisfactory two-way communication".[3] In these definitions, the key concepts are public opinion, understanding, communication and goodwill. These are the qualities a good public relations department tries to convey.

WHAT IS A PUBLIC?

Before we describe further what a public relations department does, it might be best to define what a *public* is. Actually, there are many "publics" for natural resources. Fazio and Gilbert[1] broadly defined both an internal and an external public. The internal public consists of all those who are employed by a particular agency, the department or ministry that the agency belongs to, and all of the other agencies within that department or ministry. From the lowest level to the director, a common understanding of the mission, goals and objectives of the agency and the role each person plays in that mission requires good communication among the staff and promotes smooth operations and good morale. In this regard, everyone throughout the department should be involved with public relations. It is not that everyone needs to be a public relations specialist; PR supports the rest of the agency. However, a unified and informed group of employees can be awesome, and few things can bring progress to a standstill faster than a staff that is argumentative, divisive, poorly informed or has a poor morale.

External publics are sometimes also called "stakeholders", "clients", "customers", "users", "cooperators", and a host of other names. These are the people who interact with the resources managed by the agency or with the agency itself. External publics may form communities which are defined by Fazio and Gilbert as "a social group of any size whose members abide in a specific locality". Thus, a community is geographically bounded but may contain people with a multitude of ideas. Community relations are very important. Those living around a refuge, park, forest or other facility can be considered a community, and are often those that are most directly affected, positively or adversely, by the government facility. Astute managers will make serious efforts to communicate with the communities around them, sometimes even becoming members of civic organizations within the community. Individuals within the community can include but are not limited to:

- Non-consumptive recreationists – canoeists, hikers, picnickers, and so on that do not fish, hunt, or harvest timber but enjoy being out of doors. These people see the facility as a place to get away, to enjoy nature or to recreate with minimal interference.
- Landowners who hold property adjacent to the unit – these people can be sources of problems, be bothered by unit visitors who stray off the facility, or harmonize well with the facility.
- Consumptive users – hunters, fishermen, mushroom pickers, firewood collectors, etc. Unlike non-consumptive users, these people perceive the facility as a source of food, or resources that they want to remove. This raises an assortment or regulations and behavior not observed in non-consumers.
- Business community – businessmen have a variety of unique attitudes, seeing the resource facility as an asset in attracting clients, as a source of profit through providing services to the visitors, or as a waste of good land that would be better used if it were developed into a shopping center or housing.
- Youth and their leaders, who use the facility for education and personal development.
- Media who are looking for a good story.

It is a general rule that good relations develop a sense of pride and ownership among people in neighboring communities and most formal public relations activities are aimed at these people and the broader group of external public.

Getting back to the internal public, there is one area of communication that can often be helped by good public relations but again is often ignored by the agency. For this purpose, natural resource agencies can be thought of as having three major elements (Figure 11.2) – administration, management and research – and the problem might be called the *Infernal Triangle*. Each of these elements views the agency from different perspectives. They have their own needs, responsibilities and functions to accomplish their jobs, and have different levels of education and knowledge. In a generalized way, administrators establish the goals, objectives and regulations of the agency. They should be most interested in directing and assisting other elements of

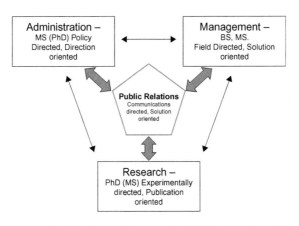

FIGURE 11.2   The Infernal Triangle. Most agencies' NGOs that have both research and management have three major categories of employees – administration, managers, and researchers. Each category has its own interests, general education level and directives. A public relations component can help make sure that the three categories communicate with each other.

the agency to accomplish those goals. Managers are field personnel, trying to accomplish conservation goals in a very practical, hands-on fashion. The common level of education for both of these groups is typically a master's degree in some natural resource-related field, but we also find many that have bachelor's degrees and perhaps a scattering of PhDs. Researchers are motivated by finding out new information through experiments and studies that may have applied applications but often have basic components. Today most principal investigators have a PhD and a vernacular that reflects their specialty. Technicians that assist research normally have master's degrees. These differences among functions in an agency can lead to disorganization and lack of communication. A manager, for instance, may desire to do her best to achieve quality wildlife habitat in a forest under her supervision, but when she turns to research to find solutions to problems she has she may perceive that the scientists are not all that responsive. It may seem to her that scientists would rather focus on tangential aspects of her needs and produce products (i.e., scientific publications) that do not really help her find solutions that relate to her problems. To the manager, scientists may seem to be more interested in designing basic studies suitable for a master's or PhD student than in trying to help her solve problems directly. The scientist is rewarded for publishing his results in journals. Unfortunately, most managers and administrators rarely read the scientific literature, and when they do they often cannot identify the solutions that they had requested. Similarly, administrators typically do not stay current with the latest scientific achievements and establish new goals for the agency without knowing if the old ones were fully addressed. As a scientist myself I have often witnessed the plight of managers but felt unable to be of much help, because what motivated them would not be appropriate for study or research from my perspective – solving their specific problems would not garner approval from my supervisors. In 2012, Gibeau[4] said that science can often be conducted in an a-contextual way, ignoring the larger social debate and political pressures at play. This knowledge gap between science, administration and management has been well-documented but the issue is still largely unsolved.[5-7]

Tackling problems in communication seems like a very appropriate task for public relations where staff that are conversant with scientific studies can take the information and repackage it into ways that administrators and managers can understand and appreciate. I once took on a task to write a couple of columns on the importance of wildlife toxicology in a magazine that catered to a wide variety of wildlife professionals.[8] Although I tried to write in a style that I thought was easy to understand, and entertaining, the assistant editor, a public relations specialist, really improved the quality of the article and made it much more readable.

Much of the rest of the chapter will deal with the external public – all of the users of natural resources.

WHAT FUNCTIONS DO PUBLIC RELATIONS DEPARTMENTS ACCOMPLISH?

Fazio and Gilbert[1] likened public relations to dental health. So often, people neglect to take good care of their teeth and gums until something really terrible like an abscessed tooth occurs and major work needs to be done to remedy the situation. If they had been proactive and gone to the dentist on a regular basis, there would be less pain and less time wasted. Agencies often neglect their public relations until problems arise, and then staff members have to be problem solvers instead of problem avoiders. Good, ongoing public relations can avoid many major catastrophes from happening.

The primary purpose of public relations is to facilitate *communication*. A general theory of communication (Figure 11.3) involves two individuals. When two people are communicating at any given time, one person is the "emitter" – the one with a message that needs to be conveyed. The message may be in coded in many ways as a signal – verbal, written, visual – and is affected by the condition of the emitter at the time. The emitter's mental state (excited, calm, angry, etc.), physical state (there may be some problem like a lisp or poor penmanship) and physiological state (hungry, sick, tired, etc.) may affect the clarity of the signal transmission. The signal has to pass through the environment, where it may be altered. Anyone trying to speak across a large convention hall will know what I mean. The other person, the "receiver", detects the signal containing the message, and the receiver's status may affect the reception of the signal so that the perceived meaning may not be the same as the intended message. As a specific example, scientific papers are often stilted in structure, with a lot of information presented very

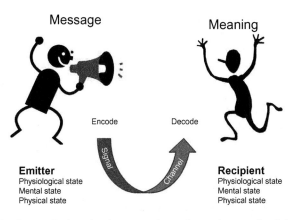

FIGURE 11.3   Communications involves an emitter who produces a signal (motion, sound, etc.) which is conveyed through the environment and reaches the recipient. Motivational and physiological states of the emitter can affect signal clarity, the environment may affect the transmission of the signal, and motivational and physiological states may affect how the recipient interprets the meaning.

tersely and words packed with specific information. The manager or administrator who is not used to this writing style may find it tedious to read and difficult to understand. Another example is that each PR program has a *focal audience* — the group of people that is intended to receive the message. Knowing the characteristics of that focal audience (its age, education level, priorities, biases, etc.) can greatly aid the PR expert in shaping a program. A job of the personal relations professional is to unpack the message into a format that is easily understandable by the intended recipients while making sure that the meaning is what was intended by the emitter. To do this effectively, the public relations professional must also know the field well and then decide how to present the message in an effective format — newsletter, fact sheet, bulletin, series of frequently asked questions (FAQs), poster, video, news releases, targeted mailings or other media — repack the message into that medium in such a way that the focal audience will understand the message, and finally disseminate it to the intended audience. Fazio and Gilbert[1] spent many pages describing how these processes are conducted, but we simply cannot go into that much detail here.

In most agencies there are several specialists within a public relations department that facilitate communication in particular ways. A large percent of government agencies and NGOs have education and interpretation functions. The staff members in education receive training on how to present information to audiences of different ages; sometimes they are certified teachers. Their specialty is to take often technical information about the agency or resources it manages and prepare it in a way that will stimulate and instruct youth and adults. They can work in a classroom, make school presentations, or work outdoors. An Office of Information, on the other hand, typically disseminates information without a lot of interpretation. If you want to know how much a fishing license costs, for example, this office might be a good source. Communications and Public Affairs offices are often synonyms for public relations. Public Involvement groups advocate the mission of the agency and work with the public, but elicit public support in actions such as cleaning up debris from roadsides or even participating in research. When studies do not require extensive training, public participation can be doubly rewarding by assisting in data gathering and encouraging goodwill between the agency and the public.[1,5]

PRINCIPLES OF PUBLIC RELATIONS

Fazio and Gilbert[1] outlined seven principles of public relations:

1. *Every action makes an impression.* If a disgruntled custodian at a visitor center complains about their pay, or the receptionist about the working conditions or "that idiotic manager", you can bet that the visitors will be affected negatively. On the other hand, if a visitor encounters a field crew that is cheerful and upbeat, or if a real, knowledgeable person answers the phone instead of a recorded message (a rare occurrence nowadays), the caller's impression of the facility will be enhanced. Recently I called an office of an NGO and the receptionist answered "Good morning, how may I help you smile today?" She already had.

2. *Good public relations is a prerequisite to success.* Good relations will lead to a smooth interface between the public and the agency, and this will lead to harmony in the agency. Abraham Lincoln knew that "Public sentiment is everything. With public sentiment nothing can fail: without it nothing can succeed." At the beginning of this chapter we discussed how "the Public" is involved with all aspects of natural resource conservation. Will a facility more likely be benefitted by receiving congratulatory letters or by negative letters? It used to be said that a dissatisfied customer will tell ten others of his disappointment but a satisfied person will tell only two or three. In today's world of social media, that saying is terribly outdated. According to one account,[9] on July 6, 2009, a Canadian musician uploaded a song onto YouTube that chronicled a real-life experience of how his Taylor guitar had been broken by poor airline handling during a trip in 2008. By the end of the first day the number of hits on that song totaled 150,000, three days later they had reached 500,000, by mid-August there had been 5 million hits, and by April 2010 that number had risen to over 10 million!

3. *The public is actually many publics.* We have already discussed this; see also Chapter 9 for various types of NGOs, each of which can be considered "a public". The natural resource manager and administrator need to

be aware of the publics that are involved in some way with the facility or agency. Various publics respond better to certain types of information transfer than others. An experienced biologist would probably not make exactly the same presentation to a hunting club and to a classroom of university students. It is an unfortunate truism that wealthy and politically connected stakeholders hold more power than the average citizen, even in a democracy. While everyone deserves respect and the right to be heard, there are some who demand greater attention than others.

Rodgers (1951)[10] relates a story that is repeated in Fazio and Gilbert[1] and abbreviated here. Bernard Fernow, one of the first directors of what would become the US Forest Service, served from 1986 to 1898, when he became the dean of the first forestry college in the United States at Cornell University in New York. The college received a large tract of forest to use as a demonstration site. Although the tract was in a relatively remote area, its neighbors included several very wealthy and prestigious citizens of New York. Fernow quickly went to work to set up a demonstration site within the forest, and the activities and noise created by their activities, as well as the smoke and odor from prescribed burns, upset the neighbors. Fernow unwisely ignored their complaints. Although the university strongly supported him, the neighbors took their complaints to the governor, who totally cut the College of Forestry out of the next year's budget. Years later the school was reinstituted – but at the University of Syracuse, *not* Cornell. Power has a strong voice.

4. *Truth and honesty are essential.* Perhaps 50 or 60 years ago the unscrupulous natural resource manager might have been able to ignore the public or persuade it with half truths or cover ups without too much difficulty. Today, the American and Canadian public are better educated; many have at least some college education and have taken courses in conservation, or they have learned about the environment through the media. Thus the public is more interested in the environment around them, more knowledgeable, and less satisfied with inadequate answers to their questions. Nowadays the general public is also very suspicious of anything that may appear to be a cover up or obfuscation. A hint of anything less than the total truth can quickly erode the respectability of an agency, and trust is very difficult to win back once it has been lost. Time and time again people in the public limelight have learned, to their dismay, that it would have been better to be forthright than to try to sweep an issue under the rug.

5. *Offense is more effective than defense.* A wise manager or administrator knows that prior preparation is critical to winning public sentiment. Media campaigns should begin *before* a major activity starts, to explain what is going to occur and the reasons for the activity. Staff should also know enough about their stakeholders to have a good idea of how the different elements will react to a planned change. They will know how non-

consumptive users might respond to increased logging in a forest, or how the hunter will respond to reduced season lengths. Agencies are recognizing this fact and are increasing their efforts to understand their publics through surveys, questionnaires and other forms of fact-finding. In fact, there is even an entire aspect of natural resources science called *Human Dimensions* that focuses on human attitudes and perceptions. Case histories that appear later in this chapter illustrate the value of this discipline.

6. *Communication is the key to good public relations.* There is no way around this one. The free flow of communication is the very heart of PR. Communication is the most important thing humans can do in any relationship, yet it is also one of the most difficult things to do right. Almost all confrontations among people can be traced to problems with communicating. Communication gaps can occur both internally within an agency, and externally between the agency and the public. The public relations professional, therefore, is always studying ways of how communication can be made more effective.

7. *Planning is essential.* Just like in other elements of natural resource administration (see Chapter 12), planning is critical in public relations. There are many details that occur in a PR campaign, from determining what the agency wants to convey, to judging how to say it, to selecting the most effective method of presenting the message. Planning keeps the agency on the offensive, one step ahead of critics. Planning also leads to more efficient communication, because the agency is aware of the perceptions of the public and can adjust to those perceptions or even shape them before they become problems.

PLANNING A PUBLIC RELATIONS PROGRAM

There are similarities between conservation plans and PR plans or strategies, in that both require taking stock of the needs of the agency, developing goals and objectives, implementing the plan and, hopefully, evaluating the success of the plan. However, since PR plans deal with persuasion and communication and not directly with wildlife species, forests or habitat, there are some elements that differ. Public relations strategies would fit nicely within some of the comprehensive plans required by federal, state and provincial governments, but most agencies have not become that forward in thinking yet.

A PR strategy consists of several steps (Figure 11.4). Fazio and Gilbert[1] described the "*5 M-s of Marketing*" that define a marketing plan, and these work as well in natural resources as they do in business.

1. Message – What is to be said (or sold)?
2. Market – Who is our audience, our public? What can we know about this group of people?

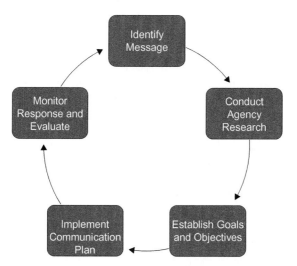

FIGURE 11.4   The planning process for a public relations strategy is similar to that of a comprehensive conservation plan for the agency. Both involve research, definition of the goals and objectives, implementation of the plan, and evaluation.

3. Medium — How are we going to disseminate the message? Will we use brochures, fact sheets, newspaper articles, purchase spots on radio or TV?

4. Money — What is this going to cost, and is it in the budget?

5. Measurement — How will we determine if the plan was successful? What outcome do we expect to see?

The first step, message, is relatively straightforward. What do you want to tell the public? In a previous chapter we discussed how the US Forest Service went from having Smokey Bear say "Only you can prevent forest fires" to "Only you can prevent wild fires". The new message was based on the Forest Service acknowledging that prescribed burns were beneficial and incorporating them into their forest management plans. However, the public, long indoctrinated that all fires were bad, needed to be persuaded and educated that some fires were actually good. When the California condor (*Gymnogyps californianus*) was removed from the wild, the US Fish and Wildlife Service recognized that it had to develop a campaign to publicize why they were taking the birds out of circulation and what they were going to do with them. Before the agency released birds back into the wild, the FWS developed another campaign to educate people about the hazards of lead shot and the importance of protecting the released birds. Several NGOs and other public groups wanted to persuade the Canadian legislators of the importance of the Species at Risk Act (SARA) during the time that the government was debating enacting the law, and these campaigns were effective in passage of the Act.[11] Once the message has been identified, the next step is research (called marketing in the PR business).

Research in PR of natural resources has several purposes. Research on the stakeholders can help identify potential issues before they occur. Research can also identify groups that can lend support to management decisions. Data gathering can be accomplished through formal surveys. It may also include having meetings between agency staff and stakeholder groups, such as NGOs, local community representatives and the general public. It is important that the technical staff – those that meet the public during the daily course of their jobs – alert public relations staff about potential issues or desires that they see or hear from their contacts. Perhaps the public is increasingly observing violations of creel limits on refuge or state lakes and reporting this to any staff member they see. Staff then should be directed to pass these observations to supervisory staff, who can initiate tighter surveillance and heighten publicity about the state's "Turn In Poachers (TIP)" program, making the public more aware that they can do something for the resources. Research can also help identify the most effective ways of reaching the public so as to assess the variety of attitudes of stakeholders that interact with the agency. This information gathering can be used to fine-tune public relations programs and can be an ongoing process so that communication can constantly be improved.

Goals and objectives are established to accomplish a desired outcome. In public relations as in other fields, *goals* tend to identify the final outcome whereas *objectives* define the steps to reach that outcome. For instance, government agencies have long recognized that the composition of their employee bases does not match that of the general public. Discrepancies exist in ethnic, gender, racial and disability categories. Therefore, governments have established a goal to make their work forces more representative of the American or Canadian general public. To accomplish that goal, specific objectives such as training and employment opportunities have been established. Public relations departments have gotten involved by making these programs known.

Implementing the communication plan first involves determining how the agency is going to spread the word – what media is it going to use? There are many ways of disseminating information, and each has its own characteristics based on financial and manpower costs, the size of the audience it can reach, its specificity for a focal audience, preparation time, and other factors. For example, a campaign based on television would likely reach a large number of people, but only a small fraction of those would have any interest in the message. At the same time, television ads are expensive and it takes a lot off effort to make a 30-second spot that will catch viewers' attention. While television might be appropriate for a brand of soda or dish detergent, it might not be an efficient method for many natural resource purposes. One way of making television or other mass media venues more efficient is through spot placement – using them in areas that have higher percentages of potentially interested people. Ads about forest fire might alert many people in heavily wooded states such as California or Colorado, but would likely have less effect in New York City. The most efficient campaigns are those that involve a few modes of communication aimed at a focal audience. For example, people coming to a visitor center might be asked to register with their names, email addresses and residential addresses. Follow-up messages via the Internet or postal service could then be sent through the year informing the visitors of upcoming events or issues of concern. The audience would be small but motivated. There are times, however, when messages need to reach larger numbers of people, so selecting the most effective method is very important. The rest of the implementation phase involves preparing the message and transmitting it through the selected media. There are too many variations in these processes to present here.

Measurement and evaluation occur once the communications have been sent out. Like many of the conservation plans (Chapter 12), this step is often neglected.[1] However, if there is no measurement or evaluation, how can an agency or NGO know if it is reaching the focal audience and if the meaning that the audience perceives is consistent with the message being sent? These steps complete the process and provide information that will be useful for subsequent media plans.

One way of assessing the effectiveness of a public relations strategy is to determine how people respond. If an agency invites people to attend an event, it is easy to count the number that actually come. Another way is to send out a survey to a portion of the audience that supposedly received the message in the first place. Surveys sent by government agencies, however, can be restricted by laws and policy. Another method is to have the audience return evidence that they received the message, such as a self-addressed post card. This method is notorious for poor return rates, but if an expected return rate can be estimated from prior experience it can then be compared to the actual rate. The bottom line of all of this is to determine whether the goals of the plan were met. If the goal was to reduce littering, is there actually evidence of reduced roadside debris following the campaign? If it was to encourage the public to report poachers, is there an increase in notifications compared to previous years?

CASE STUDIES

In this section, we take a look at some examples of how public relations benefits natural resource agencies.

Case Study 1: Know Your Public

Ranchers in certain western states or southern portions of some provinces may encounter various species of prairie dogs (*Cynomys* spp.), which are considered by many to be vermin because cattle grazing through a prairie-dog town could stumble and break a leg by stepping into a burrow, and because the small herbivores might compete for scarce grasses in arid regions. Historically, extensive poisoning, shooting and bounty programs began in the late 1880s, and by the 1920s millions of prairie dogs had been exterminated.[12] These programs still continue today locally, but by 1960 prairie dogs had been reduced to around 2% of their historic range, and two of the five species are now threatened or endangered in the United States. Prairie dogs are keystone species of the west; adults and young are food for many predators and their burrows provide cover for several other species, such as burrowing owls (*Athene cunicularia*), the endangered black-footed ferret (*Mustela nigripes*), swift foxes (*Vulpes velox*), and snakes. In summary, conservation of prairie dogs, especially those that are not considered to be of concern, such as the black-tailed prairie dog (*Cynomys ludovicianus*), is problematic.

Public attitudes towards prairie dogs were assessed in Montana as a step towards instigating a conservation program for the prairie dogs and the dependent black-footed ferret. In 1999, Reading and colleagues[13] reported on the results of a survey of 900 randomly selected people, including 300 rural residents from one county in Montana that had active and ongoing black-footed ferret management, 300 urban residents in Billings, 150 randomly selected ranchers, and 150 members of conservation NGOs. These groups were considered to be different elements of the same focal audience. Response rates ranged from 64% to 85%. As might be expected, there were major differences among the attitudes of the different groups. Conservation NGO members and urban residents were most favorable towards prairie dogs, county residents were more negative, and ranchers were very negative towards the animals. Negative attitudes focused on the pragmatic − damage to grassland habitat and loss of control over range management. Positive attitudes were more based on ethics, animal rights and ecological benefits. Only members of conservation groups were actually concerned for (not about) prairie dogs.

One question the survey asked, "How much of the public grazing lands would you like to see maintained as prairie dog colonies?", reflected this difference among publics (Figure 11.5), with a majority of ranchers wanting none and conservation group members wanting almost half of the land maintained for prairie dogs. The county group was considerably less negative than the random sample of ranchers, but not as positive as urban residents or NGO members. The differences were statistically significant ($P < 0.001$).

Another question asked where the focal groups got their information about prairie dogs. The survey found that most of the respondents received their information from personal experience, regardless of focal group (Figure 11.6). The next leading source was books and articles. While direct information from the Bureau of Land Management had the lowest response rate, television was also low. The group that had the highest response to the

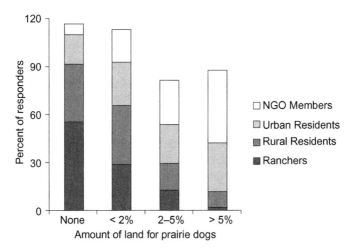

FIGURE 11.5   Percent by segment of focal audience expressing how much land they think should be dedicated to prairie dog management. *Figure created with data from Reading* et al. *1999.*[13]

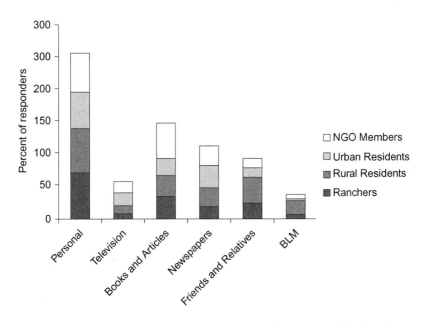

FIGURE 11.6   Sources of information on prairie dogs by segment of focal audiences. *Figure created with data from Reading* et al. *1999.*[13]

BLM, however, was the county residents, presumably because the ferret project was accompanied by a public relations campaign of some sort, even if it consisted of local gossip.

So what could a public relations department do with this information if it wanted to help conserve prairie dogs? First, staff might want to spend most of their time and effort trying to persuade groups other than the ranchers. It is obvious that ranchers have a very strong negative attitude towards prairie dogs, and that a large percentage of their knowledge comes from personal experience. Several studies have shown that those with strong negative attitudes are the most difficult to get to change their minds, and this difficulty is enhanced when the knowledge is first-hand. In other questioning, the survey found that the ranchers strongly wanted complete freedom to manage the grazing grounds without interference. There was variance among ranchers, however, so some efforts of persuasion might yield benefits, but ranchers should not be the main focus of any PR campaign. Additionally, members of conservation NGOs were mostly already on board. The public relations team may want simply to reinforce their beliefs. The urban residents had the least amount of personal experience with prairie dogs, and were generally in favor of conserving the species. Education could strengthen their beliefs. The county residents had an attitude that was similar to that of the ranchers but generally less draconian. It could be surmised that the ongoing black-footed ferret project may have awakened some interest and concern for prairie dogs in the residents compared to the general population. This is supported by this group's answer on information sources. The resident group should be a major part of the publicity campaign as this group's attitudes are changing, suggesting that it is open to further education.

So, how can the public relations people get the message out effectively? Well, television is clearly not the way. Perhaps there just was not much on TV about prairie dogs, but this is the most expensive medium to use, and unless stations are willing to give away free air time the agency might better spend its money and time elsewhere. What might be discouraging is the very low response rate to the BLM as a source of information. As the federal agency responsible for the public land, it should be conspicuous. Other than the county residents, hardly anyone is getting information from the BLM. This might be due to a general antagonism against federal agencies out west, but perhaps the public relations staff for the BLM could increase their efforts. Personal experience and friends and relatives are two sources of "information" that can be highly biased. Education programs could be helpful in increasing the accuracy of shared information. Newspapers often publish stories at no charge, and books and articles seem to be reaching some people, especially the NGO members. These venues might be seriously considered.

As Reading and colleagues note, information dissemination would not be sufficient to persuade some of these publics that prairie dog conservation is a good thing. With the ranchers' attitudes, in particular, values and behaviors would have to be addressed. While ranchers and rural residents claimed to know a lot about prairie dogs, their scores on questions about the rodents were not significantly higher than those of urban residents or NGO members. One measure that might begin to change attitudes would be to take some of the money spent on prairie dog eradication and apply it to positive incentive programs targeted towards ranchers.

The Reading *et al.* study occurred around 15 years ago. Recently, the Fish and Wildlife Service completed a status review of the black-tailed prairie dog and has determined it does not warrant protection as a threatened or endangered species under the Endangered Species Act.[14] The Service assessed potential impacts to the black-tailed prairie dog, including conversion of prairie grasslands to croplands, large-scale poisoning and sylvatic plague, and has determined that these impacts do not threatened the long-term persistence of the species. Black-tailed prairie dogs occupy approximately 2.4 million acres of range, and their estimated population in the US is approximately 24 million. However, the Utah prairie dog (*Cynomys parvidens*) is a threatened species, and the Mexican prairie dog (*C. mexicanus*) is endangered.[14]

Case Study 2: Tending to the Internal Public

As we have pointed out, the Infernal Triangle can be a serious communications problem in some agencies. In 1999, Johnston *et al.*[15] reported on a study of this problem within the US Department of Agriculture's Wildlife Services program. The overall purpose of Wildlife Services is to minimize wildlife damage in urban, agricultural and natural environments, and to reduce wildlife-related risk to human health and safety. At the time of the study, the agency had a Research component composed of about 150 scientists, technicians and support personnel headquartered at the National Wildlife Research Center, Fort Collins, CO, and nine field stations throughout the US. The other major entity within Wildlife Services was Operations, with around 800 employees throughout the United States, which provided a range of hands-on and advisory services to governments, individuals and businesses. Research focused on developing new methods to decrease wildlife/human conflicts, whereas Operations substantially implemented these methods. The study was conducted in response to frequent comments by Wildlife Services personnel that communications could be improved between Research and Operations. Many agencies consider PR as primarily being between the agency and the external public, but information transfer is a public relations activity, regardless of the audience.

Johnston and colleagues proceeded to conduct a survey of Wildlife Services personnel to determine the quality of information transfer between the two divisions and to assess how that transfer could be improved. They contacted 580 employees and 39 state directors. The response rate was 81%. Questions focused on how well staff members were acquainted with animal control methodology developed from within Wildlife Services and from external sources. The study indicated that they were receiving more information internally than from outside sources, which was good, but still 31% of the respondents did not believe that they were obtaining sufficient information internally or externally. Dissatisfaction was evenly spread throughout the agency, across geographic regions, job categories and years of service. The investigators determined that operations staff were more familiar with animal control methods that they had used than with those that were newer or had not been released. A set of questions was aimed at the most effective way of communicating new information to operations staff. Of seven methods (Internet, audio cassettes, videos, fact sheets, research articles, workshops and one-on-one opportunities), the most effective methods in the past were the same as those predicted to be the best for future information transfer: one-on-one meetings and workshops.

This type of information can be very useful to a public relations or similar department. In this case, it told the agency that while in-house information transfer was more efficient than that from external sources, there was still plenty of room for improvement. It identified certain animal control methods that needed to be better explained, and it indicated which methods of discourse would be most effective. The administration then had information that could increase the efficiency of their efforts to improve in-house, cross-division information transfer.

Case Study 3: Science and Public Relations Can Work Together

This case study takes us to Germany and butterflies.[16] I chose a German study because, unfortunately, there are not many published case histories that have clearly demonstrated a relationship between public relations and the development of a new science project in either the United States or Canada. There are many studies on stakeholder attitudes, but very few that associate a public relations strategy and the end product.

In Europe, as elsewhere, insect species constitute a huge percentage of the flora and fauna in a given area. Among insects, butterflies and moths are iconic representations of a healthy environment. Many people love to watch butterflies; they are easily seen, charismatic and diverse, and local species composition can provide insight into the environmental health of an area.[16]

Following a similar effort that started in England as early as 1976, a nationwide monitoring program, the Tagfalter-Monitoring Deutschland (TMD), was established in 2005 with a headquarters at the Helmholtz Centre for Environmental Research. An objective was to develop a cadre of trained citizen scientists to monitor the butterfly and diurnal moth populations around the country. To attract public attention and volunteers for the project, a public relations campaign entitled "*Abenteuer Schmetterling*" (Adventure Butterfly) was launched from a popular science program on German public television. The station reported on the decline of butterflies throughout

Europe, and ongoing research. Throughout the summer of 2005, volunteers were urged via the television program to become transect walkers. At the same time, Germany's chapter of the NGO Friends of the Earth encouraged its members to become butterfly recorders. Other aspects of the campaign included a contest for the best ideas for butterfly conservation in gardens, schools, and public areas, and a challenge to the public to find six different species of butterflies each day. Other television stations and newspapers picked up on the project and gave their own presentations.

The TMD program, which still existed at least as late as 2013, consisted of transect walkers, regional coordinators and central coordination. New transect walkers were asked to recommend transect sites near their homes. Regional coordinators were experts in butterfly identification and personally paid a visit to each new transect walker. The coordinators approved the transect(s), and provided information on the standard sampling protocol and species identification. They also assisted walkers with identifying unfamiliar species. Volunteers (walkers and coordinators) were kept informed of the program through newsletters, annual reports and regional meetings. Coordinators also met annually with other experts to discuss progress and problems, and the annual meeting of professional lepidopterists had a special section dedicated to volunteer monitoring.

The Helmholtz Centre for Environmental Research maintains a web page in cooperation with a publisher of interactive media, *science4you*, for data recording, and providing information and a forum for discussions. To enhance volunteer involvement, *science4you* connected TMD with other volunteer programs across Germany, all of which shared the same technical resources. Today, the program has spread to other countries, including Israel, China and Australia.[17] This project is an excellent example of how someone can develop a scientific program from a perceived need and get it rolling through a multi-faceted public relations initiative.

Case Study 4: Working with the Local Community

We mentioned that one of the very important concepts of good public relations is to work well with the local community. These communities can be large cities, several surrounding towns or very small villages, such as in this example. Many of the same public relations tactics can be used regardless of the size of the local community, although the most effective programs will incorporate the attitudes and unique features of the community.

Brier Island is an isolated village of around 250 people at the end of the Digby Neck in southwestern Nova Scotia (Figure 11.7). It is an important ecological area because many seabirds and whales migrate through this area, and it is on the edge of the Bay of Fundy with tides that exceed 15 m – among the highest in the world. The island also provides habitat for several rare plants, including the endangered mountain avens (*Guem peckii*). In

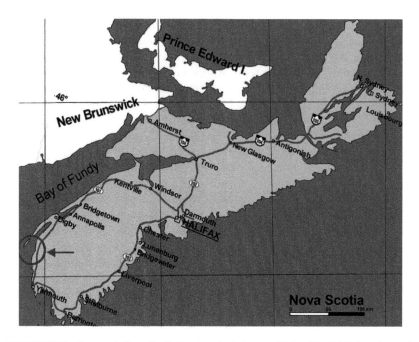

FIGURE 11.7   A map of Nova Scotia showing the isolation of Brier Island (circle and arrow) on the Digby Neck. *Source: Province of Nova Scotia.*

addition, despite its small geographic size and human population, Brier Island inhabitants have a diverse range of opinions concerning natural resources and conservation.[18]

Hennessey and Beazley[18] discussed how the attitudes of a local population can be used to leverage or facilitate conservation if those attitudes can be assessed and understood. Greater community support and engagement can lead to enhanced interconnections between ecological and cultural values. Similarly, cultural heritage can be used to increase the local recognition of a conserved landscape and thus foster cooperation in conservation efforts. In other words, if the local community can be united through common values of culture and heritage, and if these aspects can be tied to the conservation of an area, much can be done. It can be argued that small, isolated communities might be easier to unite under a common cultural heritage than a large, multi-faceted city or town. For example, many of the Brier Island inhabitants can trace their family's arrival to the founding of the community in 1783, so they should already have a well-defined sense of place and many concepts held in common. However, these concepts may be beneficial or detrimental to stewardship of a Nature Preserve. The purpose of the authors was to evaluate the variance in public opinion, and to explore how individuals and communities can be motivated to conserve ecosystems through expressions of their values. In particular, the study focused on how these values can be used to improve management of a 484-ha Brier Island Nature Preserve (BINP) owned by the Nature Conservancy of Canada.

At the time of the study, volunteer residents had participated in the management decisions and activities within the BINP for more than 20 years; however, enforcement of inappropriate use such as firewood gathering, garbage dumping, erosion, and ATV traffic had lapsed so that the preserve was being degraded. More importantly, community opinions of the preserve were changing and there was diversity in opinions on what to do with the preserve. Because of the remoteness of the island, self-policing was essential if the preserve was to be maintained. Polarization in a small community can be problematic because of the forced proximity of residents, and citizens that favored a stronger enforcement of BINP regulations were reluctant to confront violators who just happened to be their close neighbours. The role of a PR program, therefore, was to encourage the residents to be more unified in having a deeper sense of appreciation for BINP. As the authors pointed out, such problems are not uncommon near protected natural areas.

Hennessey and Beazley conducted an initial informal review of residents and, with the aid of other conservation agencies, selected a group that was knowledgeable about BINP and community attitudes for a more in-depth interview, using this selected group as spokespersons for the residents of Brier Island. Attitudes along six environmental values[19] were first assessed.

1. Aesthetic value — the importance of landscapes and species to the diversity and beauty of the planet.
2. Economic value — the role of financial incentives in the value of nature.
3. Instrumental value — the importance of nature in providing for human welfare.
4. Cultural value — the value of nature based on traditional or historic relationships between people and their surroundings.
5. Natural historic value — the value of understanding nature as a record of past ecological processes.
6. Intrinsic value — the value of nature just for its own sake.

All participants agreed that the BINP had important aesthetic values and were pleased that the preserve was there. Among those surveyed, economic values were not high priorities. Residents did not want to appreciably . enhance the reputation of the BINP for fear that, while advertising might bring in more tourists and hence more money to the community, those tourists were not necessarily wanted. Others expressed a desire to increase employment but, again, the risk of bringing in an undesired element tempered those desires. The participants thought that community sentiments towards instrumental values varied from those that occasionally picked berries or hunted, to frequent non-consumptive, non-damaging use of the site, to those that drove ATVs in the area and caused much of the erosion in the preserve. Cultural values were often combined with natural historic and instrumental values of the BINP. Because most of the families had long histories on the island, they took value in stories of their ancestors collecting bird eggs, hunting, fishing, and other activities. There was a bond between family history and the land. The participants speculated that collectively the residents may not have much appreciation for the natural historic values or intrinsic values except as they related to family history. The authors do not say, but it is likely that the residents of Brier Island did not have substantial formal education in ecology or conservation, so these values may not have had an opportunity to be well formed.

Place dependence (the identification of intrinsic values tied to an area) and *place identity* (how a person or community sees the role of an area in their culture) are important in the concept of "ownership". Do the members of a community perceive an intrinsic connection between themselves and the area? Both were evident in the Brier Island community regarding the BINP. Place dependence was evident in the prevailing attitude that the BINP was *the* place to go for berry picking, swimming, hiking, bird watching, rock collecting, firewood gathering and ATV driving. Thus, not all perceived purposes of BINP were conducive to the preservation of the area. Detrimental activities, while supported by some of the residents, also caused concern among others who feared that these activities could degrade their enjoyment of the preserve. However, residents also feared that too fierce an enforcement could lead to the loss of activities that they cherished. Participants revealed that conservation measures such as trail development for ATVs had been voluntarily placed, but they also were concerned about continuing environmental damage by ATVs. They expressed a need for leadership – the volunteers were there to help maintain the preserve, but they lacked guidance. Place identify was strong, to the point that input from those "outsiders" (people who had lived on Brier Island for decades but had moved in rather than being born into the community) had less import than that from "natives". Place identity, however, motivated the residents to enact some conservation measures on their own. Conservation projects included redeveloping a picnic ground, writing letters to get the Coast Guard to paint two existing lighthouses, and small single-family projects such as garbage removal.

Hennessey and Saunders[20] developed a management plan for the BINP that was based primarily on traditional ecological values. After the current survey, Hennessey and Beazley[18] suggested that incorporating more of the local community's value systems into a management plan might be very valuable in encouraging community buy-in and support. The residents placed less importance on ecological and intrinsic values than they did on cultural, instrumental and economic values. Aesthetic values were integral in both the management plan and community perspectives. Motivating residents by promoting place dependence and identity could be very useful in promoting a sense of ownership and appreciation for the preserve. An integrated approach involving all six values is likely to be the most effective use of resources. This approach "combines the interests of the community and the protection of island heritage features with the protection of biodiversity [and] illustrates how community capacity could be leveraged to assist with nature conservation".

Study Questions

11.1 What constraints are placed on a refuge manager, forest biologist or other field person in natural resources that may prevent that individual from doing what he or she may want to do?

11.2 What do we mean by "The Public"? How does the public affect operations within a natural resource agency or NGO?

11.3 What role does a public relations department have in an agency or NGO?

11.4 Describe the Infernal Triangle and the problems it can cause within an agency.

11.5 Describe the basic elements inherent in communication.

11.6 What are the six principles of public relations?

11.7 List and define the 5 Ms of Marketing as they might relate to a natural resources PR strategy.

11.8 As a class project, or working in groups, develop a PR strategy for some need in either your university or your state. Describe how you accomplished each step of the PR plan.

11.9 Choose one of the case histories, read the original paper on it, and be able to explain it to the class.

REFERENCES

1. Fazio JR, Gilbert DL. *Public Relations and Communications for Natural Resource Managers*. 2nd ed. Dubuque, IA: Kendall Hunt; 1986.
2. Webster's Ninth New Collegiate Dictionary.
3. Cutlip SM, Center AH. *Effective Public Relations*. Englewood Cliffs, NJ: Prentice-Hall; 1978.
4. Gibeau ML. Of bears, chess and checkers. *Wildl Prof*. 2012;Spring:62−64.
5. McKinley DC, Briggs RD, Bartuska AM. When peer-reviewed publications are not enough. Delivering science for natural resource management. *Forest Policy Econ*. 2012;21:1−11.
6. Shanley P, Lopez C. Out of the loop: why research rarely reaches policy makers and the public and what can be done. *Biotropica*. 2009;41:535−544.
7. McNie EC. Reconciling the supply of scientific information with user demands: an analysis of the problem and review of the literature. *Environ Sci Policy*. 2007;10:17−38.
8. Sparling DW, Rattner BA, Barclay JS. The toll of toxics. *Wildlife Prof*. 2010;4:25−30.
9. Asset-based marketing. Myth: The average unhappy customer will tell 10 people about the poor service he or she received. <http://www.assetbasedmarketing.com/marketing-news/myth-the-average-unhappy-customer-will-tell-10-people-about-the-poor-service-he-or-she-received.html>; 2011.

10. Rodgers AD III. *Bernard Eduard Fernow: A story of North American forestry.* Princeton, NJ: Princeton University; 1951.

11. VanNijnatten DL, Boardman R, eds. *Canadian Environmental Policy and Politics Prospects for Leadership and Innovation.* Oxford University Press, Don Mills, ON.

12. Bell WR. Death to the rodents. USDA Yearbook 1920. 1921:421-438. Reported in Reading RP, Miller BJ, Kellert SR. Values and attitudes towards prairie dogs. *Anthrozoös.* 1999;12:43−52.

13. Reading RP, Miller BJ, Kellert SR. Values and attitudes towards prairie dogs. *Anthrozoös.* 1999;12:43−52.

14. US Fish and Wildlife Service. Endangered species. Black tailed Prairie Dog. <http://www.fws.gov/mountain-prairie/species/mammals/btprairiedog/>; 2011.

15. Johnston J, Volz S, Bruce K, et al. Information transfer for wildlife management. *Wildl Soc Bull.* 1999;27:1043−1049.

16. Kühn E, Feldmann R, Harpke A, et al. Getting the public involved in butterfly conservation: lessons learned from a new monitoring scheme in Germany. *Israel J Ecol Evol.* 2008;54:89−103.

17. Helmholtz Association for German Research. Now butterflies are also being counted in China, Australia and Israel. <http://www.eurekalert.org/pub_releases/2009-08/haog-nba081409.php>; 2009.

18. Hennessey R., Beazley K. Leveraging community capacity for nature conservation in a rural island contents: Experiences from Brier Island, Canada. <http://www.tandfonline.com/doi/abs/10.1080/01426397.2012.731498#.Uer_Nm39yi0>; 2012.

19. Paterson B. Ethics for wildlife conservation: Overcoming the human−nature dualism. *Bioscience.* 2006;56:144−150.

20. Hennessey R, Saunders S. *Brier Island Nature Preserve Management Plan.* Fredericton, New Brunswick: The Nature Conservancy of Canada; 2007.

Chapter 12

The Bottom Line — Funding, Budgeting, and Planning

Money isn't everything ... but it ranks right up there with oxygen.

Rita Davenport

Money often costs too much.

Ralph Waldo Emerson

By failing to prepare, you are preparing to fail.

Benjamin Franklin

Terms to Know

Office of Management and Budget
Appropriations committees
Subcommittees on appropriations
Combined Conference Committee
Continuing resolution
Omnibus bill
Mandatory funding
Discretionary funding
Expenditure Management System
Central Agencies
Main Estimates
Standing Committees of the House of Commons
Statutory expenditures
Voted expenditures
Government Performance Modernization Act 2010
Comprehensive Conservation Plan
Mission statements
Vision statements
Comprehensive Wildlife Conservation Plan/Strategy

D.W. Sparling: Natural Resources Administration. DOI: http://dx.doi.org/10.1016/B978-0-12-404647-4.00012-X

INTRODUCTION

In this chapter we cover two different topics that go hand in hand — finances and planning. The quotations above emphasize the value of both. Money is not everything, but without it agencies cannot do very much. At the same time, money sometimes costs quite a bit in terms of any agency's personnel, time, and funding in making sure the agency obtains sufficient funding to carry out its mission and then to be sure that it spends its funding as intended. There is a truism in natural resource administration — actually, in almost any government undertaking: there is never enough money to do everything one would like to do. However, careful planning can help make the most of what money an agency does get. In this chapter we'll first cover the budget process and then mention some important sources of external funding.

Ben Franklin's quotation is also true, for a lack of planning very likely leads to inefficient agencies whose staff fail to learn from their experiences, and set themselves up to be less productive than they could be. This has been a hard lesson for natural resource agencies to learn. A few decades ago, very little planning was done. Agencies would issue mandates to their field staff without providing them with the tools and funding they needed to accomplish the mandates, or they would bounce around from one "hot" topic to another and waste agency resources. Today, planning has become a routine process at all levels of governments. The change, in part, is due to new laws that require formalized planning processes.

THE BUDGET PROCESS

Federal budget processes differ considerably between the United States and Canada due to their different forms of government. The distinct separation of powers in the United States between the Executive and Legislative branches of government, and a strong bicameral legislature, are supposed to establish checks and counterchecks through the budget process, but they can also set up political tension between and within branches. Both branches share about equal responsibility for the budget. However, the parliamentarian government of Canada blends the authority of the Legislative (House of Commons) and Executive (Cabinet) branches (review Chapter 5) and gives the Cabinet considerably more responsibility in forming the budget over the House of Commons. Although Canada has a senate, this body has little function in formulating the budget. In each country, states or provinces have essentially the same structure and budget process as their federal counterparts; while there may be some differences among states or provinces, it is not worth our time to deal with these trivialities. For the most part, the descriptions for the federal governments pertain to these smaller units.

The Budget Process in the United States

The start of a budget begins two to three years before it is finally passed by Congress and the President. An early phase of budget development is for the agencies to solicit and collect needs from individual cost centers within their jurisdiction. Cost centers are units such as refuges, national forests, research centers or complexes, and regional offices that are required to develop an annual budget and submit it for approval to a higher level. Staff at cost centers meet with supervisors to discuss the needs that they may have in the next three to five years. From the field level, budgets pass up to the regional offices and from there to the agency's national headquarters, and then to the department level. There are many ways that this can occur. Sometimes the budget requests are passed up linearly from the field level to regional to national levels. At other times, an agency may develop a specific initiative that affects one or several cost centers. To put the initiative forward into the process, the agency may solicit input from cost centers on what they would need to participate in the initiative. At other times Congress may establish its own initiatives with funding and pass these to specific agencies, which then parcel out the tasks and finances to cost centers. Federal research studies may be funded in yet a different way. Headquarters may decide to develop a new research initiative, or be given one by Congress. Instead of assigning the distribution of funding, the agency will initiate a call for research proposals. The call could be open only to researchers in the agency, or to external researchers as well. Interested scientists submit proposals complete with budgets, and funding is awarded competitively to the best proposals.

In a typical budget cycle (Figure 12.1), cost centers prepare their budget requests and send them to their regional offices. Regional offices review these center requests, perhaps make changes, compile them, and send them on to the agency headquarters as a combined budget from the region. Headquarters again reviews the budgets from the regions, makes changes, collates them, and prepares a preliminary agency budget. Then the agency (we are now working at the level of the US Fish and Wildlife Service, for example) sends this combined budget to the next level, which, in our example, would be the Department of the Interior (DOI). We are now at the first box in Figure 12.1, and it may have taken two years to get here as we enter the process for next year's budget. The Department budget staff work with those from the *Office of Management and Budget* (OMB) to prepare a complete draft of all of the agencies within the DOI; the other departments do the same so that cohesive budgets can be submitted. This may be somewhere between October and December of the budget cycle. OMB extensively reviews all of the budget requests to make sure that they are consistent with presidential policies and directives; they may slash or delete some requests, or add others by order of the President, but eventually it compiles the President's Budget Request, which law says is due in Congress on or before the first Monday in February.

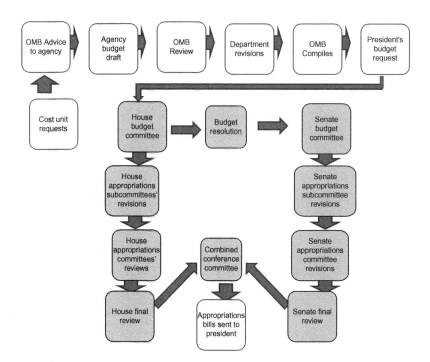

FIGURE 12.1   A typical budget process in the US federal government. Most states follow a similar plan.

Occasionally, the entire process of amassing the President's budget can be futile. If the majority in Congress is of a different party than the President, the budget may be extensively revamped. This happens somewhat regularly as the incumbent President is on his way out. In 2005, for example, the Executive Branch had to go through all of this by law, but President Bush's budget request was considered to be "dead on arrival" at a Democrat-led Congress.[1]

According to the US Constitution, all bills dealing with revenue must first be submitted by the House of Representatives. Therefore, the President's budget request first goes to the House but then is almost immediately passed to the Senate. The budget becomes separate House and Senate bills that are passed to appropriations committees in both houses. The appropriations committees send elements of the bills to subcommittees charged with oversight and budgetary matters of specific offices in the Executive Branch. The *House Subcommittee on Interior, Environment, and Related Agencies*, for example, has oversight on all of the agencies within the DOI except for the Bureau of Reclamation; the US Environmental Protection Agency; national heritage sites; the US Forest Service; the Smithsonian Institute; and about 18 other agencies.[2] The revisions made by the respective subcommittees go back to the full Senate and House Appropriations Committees for more revisions. During the subcommittee and appropriation committee revisions, substantial add-ons or removals can occur.

Most often, the Senate and House versions of the bill have differences. These are resolved in the *Combined Conference Committee*. The revised bill goes back to the full House and Senate for ratification or more debate and revisions, and this process can be lengthy. Finally, both chambers vote on the combined bill. If it fails to pass either chamber, it cycles back to the combined committee for revisions until it passes both houses. Once the bill is passed by Congress, it is sent to the President in the form of 12 appropriations bills. The President has the authority to veto, approve or refuse to sign any of the appropriations bills. If an appropriations bill is approved it becomes law, and agencies under that bill know what they will receive in the coming year. If the President vetoes a bill it has to go back to Congress, and if he refuse to sign it the bill becomes law but without the President's support. By law, the Appropriations Committee of the House of Representatives is supposed to approve the bills and send them on by the beginning of the fiscal year (October 1). Often, however, the budget becomes highly politicized and that deadline is not met, or the President sends an appropriations bill back to Congress after 1 October. When that happens, either a *continuing resolution* or an *omnibus bill* may be passed. A continuing resolution is when Congress declares that the government will use the same budget it had in the previous year, sometimes with small reductions. Continuing resolutions are enacted for a specified period of time, but may be repassed several times. In 2012, the government existed on a continuing resolution for the entire year. An omnibus bill occurs when a block of appropriations is agreed upon and passed while other appropriation bills continue to be debated. In early 1996, and again in 2013, the federal government actually shut down because appropriations bills could not be passed.

There is an important distinction in this process between budget figures and appropriations. Budgets are requests – they are not fixed estimates and, as we have seen, can be manipulated in many ways. Appropriations are actual monetary amounts that are approved by Congress and cap what an agency can spend. Agencies can spend less than what they are allocated, but that hardly ever occurs. Final appropriations can differ from budget requests. A graph for the Natural Resources Conservation Service (Figure 12.2) shows that, for this agency, appropriated funds tended to be slightly lower than requests.

Another important distinction to make is the difference between *mandatory funding* and *discretionary funding*. Mandatory funding is money that has been pre-obligated and *must* be spent by the agency. Discretionary funding is money that the agency can use for projects and operations. Over 50% of the total federal government's annual budget is tied up in mandatory funding, mostly with what are called entitlement programs (Figure 12.3A). These include Social Security, Medicare, Medicaid, income and food security,

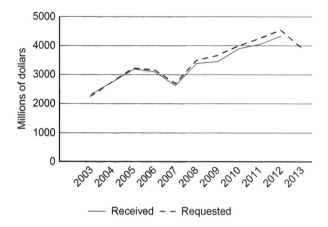

— Received   – – Requested

FIGURE 12.2   Budgets from the President provide requested funds, whereas the Appropriations from Congress are actual dollars an agency is able to spend each year. Usually these two values are not very different, as can be seen in this side-by-side comparison for the Natural Resources Conservation Service.

health care (prior to implementation of the Affordable Health Care Act), and veterans' programs. If we include the Department of Defense, 70% of the total federal budget is obligated, leaving only 30% for all other federal activities. These mandatory programs are very important, and provide compensation to millions of people. The proportion of the total US budget tied up in mandatory funding, not including military, has increased from around 25% in 1962 to just over 50% in 2012, and is expected to top 60% by 2022.[3] Conservation is not a major expenditure in this budget; even when mandatory funding is omitted, less than 4% of the discretionary budget goes into all forms of conservation (Figure 12.3B). Whether this is appropriate makes for a great debating point.

The discretionary and mandatory funding for major conservation agencies in the US federal government vary considerably among agencies (Figure 12.4). For example, the NRCS has a majority of its funding in programs tied to conservation elements in the Farm Bill. In contrast, the Bureau of Land Management and US Geological Survey have few mandatory obligations in their programs.

Also in this graph are revenues that the agencies collect from outside sources. These include grazing and other user fees, visitor fees, federal Duck Stamp revenues (discussed in Chapter 6), and similar funding sources. As a specific example, recently the National Park Service received $183 million in recreational fees, $68.4 million in concessioner payments, $22.8 million in housing rentals, and $30 million in donations and several other sources for a total revenue input of around $384 million.[4] Not all of this income can be used by the agencies to support themselves, however; some of it must be

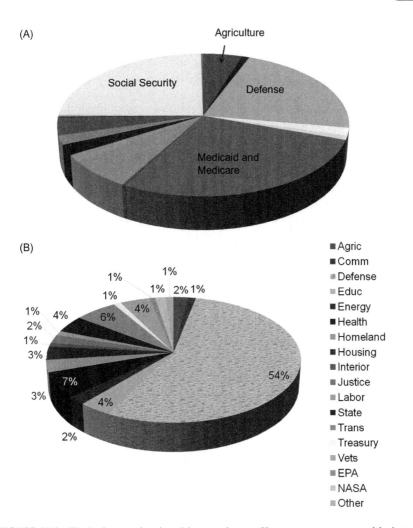

FIGURE 12.3 The budget can be viewed in several ways. Here we see a recent total budget for the United States (A) and the portion of the budget that is only discretionary (B). Social Security, Medicare, Farm Support and other programs take up a large portion of the budget, but the Defense Department's budget is huge in either category. *This figure is reproduced in color in the color plate section.*

returned to Congress for the general fund. One other aspect of the federal budget process to be discussed is *Full Time Equivalents* (FTEs). An FTE equals one 40-hour per week employee for a year. It may consist of one full-time person or two or more part-time people whose total work hours amount to 40 per week; seasonal help may also be employed for a portion of the year and count for a fraction of an FTE. FTEs are budgeted each year just like dollars. In general, the allocation of FTEs has been stable or slightly increasing through the past several years for most conservation agencies.

The Budget Process in Canada

Budgeting in the Canadian federal government is more linear than that in the United States, as most of the work is done by the Cabinet.[5,6] The process is called the *Expenditure Management System* (Figure 12.5). Like the process in the United States, the budget can be a highly politicized activity in Canada. It starts in June of each year, when the Cabinet convenes to entertain broad aspects of the budget. Cabinet officials examine budget needs relative to politics, economic conditions, public issues, investments and, of course, government priorities. From this meeting the Cabinet prepares general guidelines on what the budget should emphasize, and this *Cabinet Decision* goes to the *Central Agencies.* The Central Agencies include the Privy Council Office, Department of Finance, and Treasury Board Secretariat. The Central Agencies meet with other departments in the Executive Branch to further ascertain needs. Together, the Treasury Board

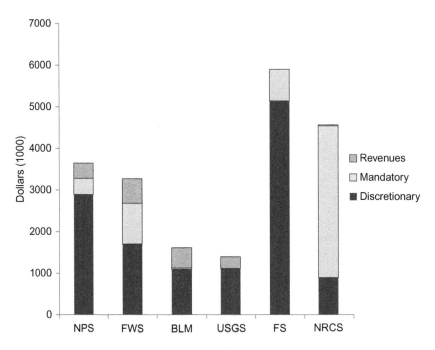

FIGURE 12.4   A comparison of recent budgets for US natural resource-related agencies. The values include discretionary and mandatory expenditures and revenues brought in by the agency for services, licenses, special taxes, and fees.

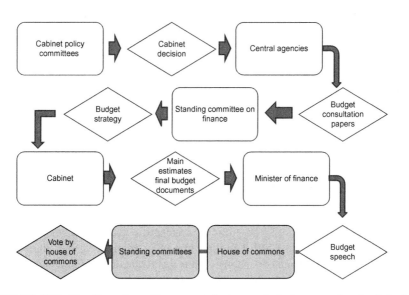

FIGURE 12.5 The Expenditure Management System or budget process for the federal Canadian government.

and parliamentary committees develop the *Budget Consultation Papers*, which cover economic and fiscal perspectives leading to spending targets. The Finance Minister releases these papers in October to initiate further consultations among the *Standing Committee on Finance* in the parliament, general public, provincial finance ministers and other stakeholders. The Finance Minister then takes all of this input and assembles it into a formal *Budget Strategy*. Between January and March of the next year, the budget strategy goes back to the Cabinet for more specific input on spending targets, initiatives and reductions. These more concrete budget estimates go back to the Finance Minister, who finalizes them. The Treasury Board Secretariat, in consultation with other members of the Cabinet, prepares the *Main Estimates* – documents that list the estimated maximum appropriations for the federal government. In February of each year the Finance Minister delivers the *Budget Speech*, which introduces the almost completed budget to the House of Commons and the public. The Estimates are then tabled by the House of Commons, thus officially acknowledging the receipt of what might be called the Prime Minister's budget; they then go to the *Standing Committees of the House of Commons* for discussion, debate and final voting. This is a serious phase for the Prime Minister, because if the Main Estimates are poorly received it could lead to unrest among the House and a change in government. By May, the individual Standing Committees report back to the full House for final debates and vote. This must come by the beginning of April, the start of a new fiscal year.

Canada has *statutory expenditures* (comparable to mandatory funding) and *voted expenditures* (discretionary funding). In contrast to the United States, however, statutory expenditures substantially exceed voted expenditures. For example, in a recent year, voted expenditures totaled C$91.9 billion and statutory expenditures were C$159.95 billion, or 63% of the total budget. While the federal budget has steadily inclined, the percent of the budget that is statutory has remained around 64% for at least the past 10 years (Figure 12.6).

As in the United States, Canadian agencies obtain funding through charges for various services, user fees, licenses, and consultations. Some of these receipts can be spent by the agency itself, whereas others go back into the general revenue stream of the country and are spent elsewhere. For example, recently Environment Canada obtained C$38.2 million for weather information services, C$16.8 million for scientific and professional services, and a total of around C$81.4 million in annual receipts. In that same year, NRCan's major spendable income came from functions dealing with clean energy (C$13.3 million), but it also received a whopping C$1.6 billion for promoting natural resource-related economic opportunities in the country; these latter receipts were not available to the agency. Compared to other natural resource agencies in Canada, NRCan has the largest statutory

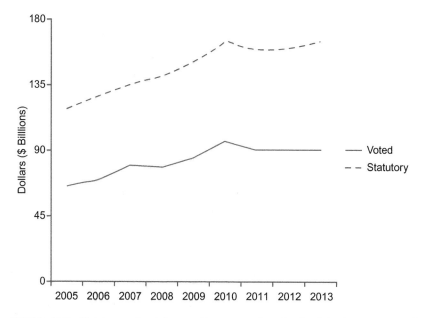

FIGURE 12.6 Statutory and voted expenditures over time for the federal government of Canada.

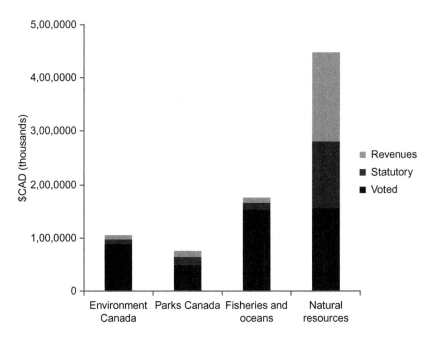

FIGURE 12.7   A comparison of voted and statutory expenditures for the principal natural resource agencies in Canada; also shown are the revenues, or receipts brought in.

expenditures and the largest revenues (Figure 12.7). Some of the largest sectors for statutory funding are investments made by the agency into various natural resource-related businesses and the Canadian economy, such as timber reserves or oil development. In turn, NRCan receives royalties and license fees from these endeavors.[7]

PLANNING

Planning in the United States

Mandatory planning is a relatively new requirement in the US federal government. Certainly, site managers have developed their own plans for the areas that they managed for decades, but the *National Environmental Policy Act* (NEPA, see Chapter 4) required that federal agencies file either *Environmental Assessments* or *Environmental Impact Statements* for any activity that involves the environment. Environmental Assessments are not as detailed as Environmental Impact Statements, but both necessitated that agencies consider what they are going to do and to plan accordingly. However, more detailed planning exercises for agencies were mandated by *the Government Performance and Results Act* (GPRA, 1993) and the

Government Performance and Results Modernization Act (GPRMA, 2010). Because the GPRMA primarily formalized the planning process, clarified some topics, and mandated quarterly reports that were originally in the original Act, I will refer to both Acts, collectively, as GPRA. Subsequently, planning and routine evaluations of those plans have become integral activities in federal agencies.

The goals of the GPRA laws are to[8]:

- hold federal agencies accountable for achieving program results
- improve program performance through setting well-defined goals, measuring program performance against those goals, and reporting publicly on their progress
- improve federal effectiveness and public accountability by promoting a new focus on results, service quality, and customer satisfaction
- help the federal plan for meeting program objectives, and provide the public with information on program results and service, and
- facilitate congressional decision-making by providing objective information on meeting objectives, and on the relative effectiveness and efficiency of federal programs and spending.

Crucial elements of the GPRA are that each agency has to develop long-term, five-year strategic plans, and that the strategic plans have to be updated and revised every three years. The strategic plans include a mission statement that defines the goals and objectives of the agency and a performance plan on how the agency expects to reach those goals. Chapters 6 and 7 include the mission statements for each natural resource agency. Annual plans that develop the goals and objectives for the current year within the guidelines of the strategic plans were also mandated by GPRA. Both Acts stipulate that an annual review of an agency's accomplishments accompanies its budget requests and is submitted to OPM and Congress.

Most agencies have other laws or regulations that take the planning process to the separate units within their jurisdiction. For instance, the National Wildlife Improvement Act of 1997 requires that all refuges and districts within the US Fish and Wildlife Service develop a 15-year *Comprehensive Conservation Plan*[9] (CCP). Like the five-year agency plans, CCPs develop long-term goals, objectives and strategies for the management of the refuge or district. These CCPs have to be consistent with the agency goals. The primary goals for refuges, for example, are wildlife enhancement, habitat management and visitor services. Any other activities that may be proposed by stakeholders or the general public during the planning process are of secondary importance, but may be implemented if they are deemed compatible with the refuge and agency goals. The refuge manager has the final say in what is compatible with the refuge mission.

Contrary to 5-year agency plan, these 15-year plans must be open to public input. Due to NEPA and the Wildlife Improvement Act, the planning team has to open the process to all stakeholders and invite their ideas, and is required to respond to all input provided. A typical planning process looks somewhat like the diagram in Figure 12.8. The first step seen at the top of the circle is to evaluate the condition of the unit. What are the major assets of the refuge? How about the major problems? Why was it established in the first place – are those reasons still valid? How do we see ourselves, and where do we want to be in 15 years? Early in the process, public opinion should be invited. Does the public view the uses and purposes of the refuge in the same way as the agency sees them? What do they want to change, remove or add? Are these desires compatible with the concept of a refuge? Can the refuge obtain authorization from the agency to implement these changes?

Taking the information and ideas from various quarters, the next step is to draft vision and mission statements. A *vision statement* is usually a single-sentence, succinct and easily remembered statement that summarizes the unit; it is like a slogan. For example, "Caring for the land and serving people" is the vision statement for the US Forest Service. A specific mission

FIGURE 12.8   An example of the steps taken in a typical planning process. There are many modifications of this, but the diagram illustrates the major elements.

statement is also desirable. *Mission statements* relate the purpose of the unit. The following mission statement provides an example[10]:

Crab Orchard National Wildlife Refuge is managed with four broad objectives: wildlife management, agriculture, recreation, and industry. The primary wildlife management objective is to satisfy the food and resting needs of wintering Canada geese and other wildlife. This objective is coordinated with the agriculture objective through cooperative farming and permitted grazing programs.

Both the vision and mission statements help to keep the goals and objectives of the refuge in focus.

After the planning team decides what it has and what the functions of the refuge are, they can progress to the next step: setting priorities. Funding and work force are always going to be limited, and it may not be possible for all of the perceived needs to be accomplished during one planning cycle, so it is critical to determine which are the most important things needing attention. Also, it must be assumed that some aspects of the plan, no matter how carefully considered, will not be accomplishable. A hundred things can hinder even the best-laid plans, so astute managers will prepare for the unexpected by designing alternatives, and alternatives to alternatives. This is one area where planning is really valuable. Instead of being reactionary, a great plan can be proactive. OK, so now the team knows what they have and what needs the greatest attention, and it has developed subplans. The next steps include putting all of this down into a comprehensive, understandable way, presenting it to the stakeholders, including those at higher supervisory levels, and preparing the final plan. Next comes that part that most biologists would consider the "fun" component – implementing the plan over the next 12 to 15 years. During this time managers are evaluating the outcomes of the plan – what was successful and what was less so than expected. Adjustments can and should be made to the plan as experience dictates. As Publilius Syrus, a Syrian writer of maxims in the 1st century BCE, said: "It is a bad plan that admits of no modification." Every three to five years the overall plan should be comprehensively re-examined. Are we still in line with the plan? Is it doing what we thought it would? Do we need to extensively revise it? After a few of these more comprehensive reviews, it is time to start all over again. As mentioned below, laws require all federal agencies to develop plans. While specific details about their plans may differ, they all require assessment, planning, implementation, evaluation and revision. They all also require the opportunity for input from multiple stakeholders including the general public.

The US Forest Service has the *National Forest System Land and Resource Management Planning Act* 1974 to govern its planning process. This Act follows guidelines set by other Acts such as the *National Forest Management Act* 1976 and *Multiple Use and Sustained Yield Act* 1979, and has been amended several times. It too requires a 15-year plan for every

National Forest. For the National Park Service, the critical document is called the *General Management Plan* and is dictated by the *National Park and Recreation Act* of 1978. Currently each park has to file a *Foundation for Park Planning and Management Plan* that relates what is unique about the park and what its specific mission is. The Foundation's document can set the stage for a more detailed General Management Plan, or serve as a freestanding document. Because each park has a unique mission, the format for planning is somewhat less formalized than in the FWS or FS, and can be for 15 or 20 years. BLM field offices or state director offices develop *Resource Management Plans* periodically for regions that have similar landscape features and conservation issues.

In 2008 the US Fish and Wildlife Service was given authority over the *State Wildlife Grant program*, which provides conservation funds to states. The program is broad and includes activities that support or improve wildlife populations, including species that are not harvested. One of the most important requirements of the program is that a state has to develop a *Comprehensive Wildlife Conservation Plan/Strategy* before it can receive funding. Prior to the federal requirement, few states had any similar plan – a lot of management was conducted on an *ad hoc* basis. These plans are to assess the health of each state's wildlife populations and habitats, identify the problems they face, and outline the actions that are needed to conserve them over the long term. The granting program provides cost-share assistance for the planning process itself. Since the plan was developed by the state and approved by Fish and Wildlife Service, the Service has continued to provide cost-sharing funding for needs identified in the Comprehensive Plans. Today, all states have accepted plans.

Planning in Canada

Canadian planning requirements are similar to those in the United States. Parks Canada has a requirement from the *Canada National Parks Act* 2000 to develop a conservation plan for each park every five years. The process is similar to the one that the US FWS requires for its wildlife refuges.[11] Because the provinces have most of the responsibility for natural resources within their borders, each has developed planning processes for forests, wildlife parks and other natural resources. *The Crown Forest Sustainability Act* 1994 (CFSA) and the *Environmental Assessment Act* 2012 (EA Act) provide the legislative framework for 10-year *Forest Management Plans* on Crown land in the provinces. These management plans are prepared in two phases. The first phase is comprehensive for the full 10 years and provides for the sustainability of the forest with regard for plant and animal life, water, soil, air and social and economic values, including recreational values. Phase II provides detailed planning for the second five years, and is developed during the last two years of the first five-year term.[12] Throughout the two-year development process for each phase, public input is actively solicited and, when possible, incorporated into the plan. Adaptive management is used to evaluate the progress and effectiveness of the plan.

These are just a few of the many plans that natural resource managers develop and use regularly. There are also plans for restoring population health and numbers for particular species, such as the *Woodland Caribou Conservation Plan*[13] in Ontario, or the hundreds of recovery plans for endangered and threatened species in both the United States and Canada. I guess my advice to anyone seeking employment within a federal, state or provincial agency is: prepare to plan.

Study Questions

12.1 Trace the major steps in the budget process of either the United States or Canada. What are the major offices involved with either process? What happens in the United States if a budget is not passed right away?

12.2 In addition to the budget appropriations, what other ways can agencies obtain funding?

12.3 How do mandatory (statutory) and discretionary (voted) expenditures differ from each other? What are some of the largest mandatory budget items?

12.4 Why is planning so important in government agencies?

12.5 Outline a typical planning process. Why is evaluation an important part of this process?

REFERENCES

1. PBS News Hour. Congress Debates President Bush's Budget Proposal for a Second Straight Day. <http://www.pbs.org/newshour/bb/government_programs/jan-june05/budget_2-8.html>; 2005.

2. US House of Representatives House Appropriations Committee Subcommittee on Interior, Environment and Related Agencies. <http://appropriations.house.gov/about/jurisdiction/interiorenvironment.htm>.

3. Austin DA, Levit MR. Mandatory spending since 1962. *Congressional Research Service.* 2012;:7−5700:<http://www.crs.gov>

4. Budget Justifications and Performance Information, Fiscal Year. National Park Service. <http://www.nps.gov/aboutus/upload/FY_2012_greenbook.pdf>; 2012.

5. Wildlife Society. Date unknown. Canada: Federal Budget Process. <http://www.wildlife.org/subunits/policy-toolkit/canada-budget>.

6. Amelita AA. An overview of the Canadian budget process. Parliamentary Center. <http://www.parlcent.org/en/wp-content/uploads/2011/04/articles_and_papers/Canadian_Budget_Process_EN.pdf>; 2005.

7. Natural Resources Canada. Report on Plans and Priorities. http://publications.gc.ca/collections/collection_2011/sct-tbs/BT31-2-2012-III-13-eng.pdf>; 2011−2012.

8. Office of the Management of the Budget. Government Performance and Results Act <http://www.whitehouse.gov/omb/mgmt-gpra/gplaw2m#h2>; 1993.

9. US Fish and Wildlife Service. Division of Refuge Planning. FWS Planning Overview. <http://www.fws.gov/mountain-prairie/planning/overview/index.html#ccp>; 2012.

10. Crab Orchard National Wildlife Refuge. <http://www.fws.gov/refuges/profiles/index.cfm?
    id = 33610>.
11. Parks Canada. Fundy National Park: A new management plan. <http://www.pc.gc.ca/pn-
    np/nb/fundy/plan/plan01.aspx>; 2011.
12. Ontario Ministry of Natural Resources. Province of Ontario's Forest Management Plans.
    <http://www.mnr.gov.on.ca/en/Business/Forests/2ColumnSubPage/STEL02_163549.html>;
    2010.
13. Ontario Ministry of Natural Resources. Woodland Caribou Conservation Plan. <http://www.
    mnr.gov.on.ca/en/Business/Species/2ColumnSubPage/MNR_SAR_CRBOU_CNSRV_PLAN_
    EN.html>; 2013.

Chapter 13

What Next?

Here we are at the final chapter in this book. We have covered a lot of material on natural resources, especially the renewable kind. Just as a brief review, we started by discussing what natural resources are, why it is important to conserve them, and different concepts of conservation, including preservation, sustainability, ecosystem services and values. From there we discussed the history of conservation, especially wildlife, in North America. While several mistakes were made along the way, the United States and Canada are better off environmentally than they were even 50−75 years ago. We still need to be concerned, but we have the knowhow and the technology to help us get through future crises. We have seen that an essential aspect of the history in both countries was the enactment of laws and establishment of legal precedents that gave us a foundation for protecting our natural resources now and in the future. The long course of history since the first colonists until now has led to the Public Trust concept and the North American Model of Wildlife Conservation in both the America and Canada, which upholds democratic principles in making outdoor recreation, hunting and fishing available to all, not just the select wealthy.

Throughout this history both countries have developed an infrastructure of governments to protect us from ourselves. Garret Hardin was not far from the mark when he explained that human nature drives us to improve our lot. Even if we are not consciously intent on abusing resources or damaging the environment for everyone else, our personal drive can and does affect others. Therefore, society establishes rules and regulations and the necessary bodies, such as courts, legislatures and authority, to enforce them. Enforcement first assures that everyone has a share of the pot, and secondly makes certain that resources will be available to future generations.

Generally people in the United States and in Canada have the same wide breadth of interests, ranging from those who place the highest priority on the environment to those whose focus is strongly on the financial profit margin. It is hard to imagine that anyone would be against a healthy environment, but some do not consider it a high priority. Agencies at the federal, provincial and state levels have the task of making sure that conservation, the "wise use" of natural resources, is followed by all. Both nations have this same goal, although differences in their governmental structures can be seen in how they go about accomplishing it. Along the way the government

D.W. Sparling: Natural Resources Administration. DOI: http://dx.doi.org/10.1016/B978-0-12-404647-4.00013-1
319

agencies interact with non-governmental organizations with their myriad of special interests and perspectives.

Finally, driving all of this is society, people. People create most of the problems concerning the conservation of natural resources, but people also provide or attempt to provide the solutions to these and other problems. As we have seen, the "public" takes many forms. It is critical for agencies to develop and maintain good relations with their publics, for the taxpayers provide the money that allows the agencies to function and have the power, through their elected officials, to help direct an agency in establishing goals and programs.

So, we have discussed where the nations have been and what exists at this time. I would like to spend a few words discussing what I believe are important goals and issues for the future. All of these have their roots in the present but are likely to become even more of an issue in the future.

1. *Global Climate Change*. This is the current environmental "hot button". It is based on scientific data, seems to be manifesting itself in various ways, promises to become worse, is accepted by a large majority of scientists, and is commonly ignored by the general public. According to a Pew survey[1] conducted in March 2013, public opinion lags behind the scientific conclusion, with only 40% of Americans and 54% of Canadians recognizing global climate change as a matter of major concern. A primary reason for the low "buy-in" into global climate change, I believe, is that it is a very complex subject with many seemingly contradictory predictions. While some areas of the Earth may become warmer, others may cool down a bit. In some areas global climate change will manifest itself more in an increase in variability and intensity of storms than an actual warming. Even the confusion of the amendment from global climate warming to global climate change has caused people to be skeptical of "environmentalist doomsday predictions".[2] A lot of the predictions are based on models that are nearly impossible for the average citizen to understand. The global climate community has seen some major scandals associated with emails that have put some scientists in a bad light and added to the skepticism. Finally, interpretations of data that have appeared in peer-reviewed journals have been questioned because a major global climate change has to be viewed from the perspective of geological time, and accurate weather recording is only a little blip on that timeline. The issue of how strong human or anthropogenic causes are is another point of debate, even among scientists. There is also evidence that the primary media sources, in an apparent effort to "present both sides" of the global climate issue, have fueled this skepticism among the public, thereby delaying any action by the governments.[3]

To the point that anthropogenic activities are causing global climate change, there is general consensus that so-called "greenhouse gases", particularly carbon dioxide from energy generation, are the primary contributors towards warming. It is great that the United States and Canada are both taking measures to reduce their generation of these gases internally. However, since climate change is global, global efforts are needed to be most effective in reducing the risks of this threat. In 1997 the United Nations convened international meetings on reducing greenhouse emissions in Japan that led to the Kyoto Protocol. Canada was active in the negotiations that led to the mandatory provisions in the protocol, and signed the Accord or international agreement that included this protocol. However, greenhouse gas emissions continued to increase in the country, and in 2011 it withdrew from the Accord. Canada did sign the Copenhagen Accord to reduce greenhouse gases in 2009, but that agreement is non-binding. The United States refused to sign the Kyoto Accord, citing an inherent bias against developed countries. The United States was involved in drafting the Copenhagen Accord and is trying to adhere to its provisions. Both Canada and the United States, as major world powers, should take the lead in drafting an international greenhouse emissions reduction plan that is fair to all nations and has a real chance of reducing emissions before some of the dire predictions of global climate change take effect.

2. *Massive Extinction Rate.* Wake and Vredenburg[4] raised the question as to whether the world is in the midst of the sixth great extinction event in history. Five other major extinction events occurred in geological time: (1) ~439 million years ago (Mya), when 25% of the families and nearly 60% of the genera of marine organisms at that time were lost; (2) ~364 Mya, when 22% of the marine families and 57% of the genera disappeared; (3) the huge Permian−Triassic extinction event (~251 Mya), when 95% of marine and terrestrial species died out; (4) the end of the Triassic (~199−214 Mya), when massive volcanic action and lava floods caused considerable global warming and loss of around 22% of the marine species; and (5) ~65 Mya at the Cretaceous/Tertiary boundary, when 16% of families, 47% of genera, and the dinosaurs died out.

About 1 Mya a series of climatic oscillations, including alternating periods of glaciations and melting, began; the current period of warming is but the most current oscillation. Many scientists believe that we are now in a massive period of extinctions due in part to climate change, contaminants, and other human activities such as habitat fragmentation and destruction. Of 52,667 animal species considered by the IUCN Red List,[5] 709 are extinct and 11,124 (21%) are species of concern, ranging from Vulnerable to Critically Endangered. The situation is even worse for plants; of 17,604 plant species in the Red List, 9,977 (57%) are of concern. While both Canada and the United States have endangered species programs, extinction is a global problem and both of these countries can play essential roles in encouraging international programs to protect species of concern. Certainly the IUCN and CITES are important, but I contend that further regulation is critical for saving the world's species.

3. *Sustainability.* We discussed sustainability and ecosystem services in Chapter 1. Since it is unlikely that the world is going to embrace the measure

that arguably would do the most to protect biodiversity— the elimination of humans from the scene (he says, tongue in cheek!) — agencies need to continue to develop ways of enhancing conservation measures that incorporate both human and ecological needs. Among the most important needs in this regard is assuring an adequate supply of clean freshwater. Freshwater is found in the atmosphere, soil, underground aquifers, glaciers and surface waters. About 3% of the total water balance on Earth is fresh (i.e., <1% salt). Of that 3%, only one-tenth, or 0.3%, is potentially usable surface water.[6] Globally, clean, unpolluted freshwater accounts for only about 0.003% of the total water supply on the planet. This surface water is essential for all life, and its distribution across the planet is far from uniform. Large deserts where annual rainfall is less than 10" (25 cm) are found on all the continents, while the wettest land in the world, Mount Waialeale on the island of Kauai, Hawaii, receives an average of 452" (~11,500 cm) per year. Water shortages are already a major concern in the arid southwestern United States,[7] and global climate change is likely to alter water availability for humans and ecosystems.[8] As we have seen, sustainability involves a lot more resources than water, but since we have not discussed water very much in other parts of the book, this appears to be a good place to do so. Government agencies need to continue to protect our water supplies through regulations and enforcement of water criteria standards.

Another element of sustainability that we have not discussed in any great detail except in application to endangered species is invasive species. An invasive species is one that is not native to an area but has been introduced through human activities and which threatens native plants, animals or habitats. Hundreds of plants and animals have been introduced into new areas over the past decades. A few introduced species, such as ring-necked pheasants (*Phasianus colchicus*), Hungarian partridge (*Perdix perdix*), chukars (*Alectoris chukar*), honey bees (*Apis mellifera*), and certain ornamental trees and shrubs, have been beneficial to humans without damaging the environment. However, most have been either benign or detrimental. Agencies have implemented eradication programs, and these need to be supported by the public.

4. Finally, there is one internal issue that seems to be intrinsic to most agencies, *human equality*. As mentioned in Chapter 10, women and minorities are greatly under-represented in professional leadership positions. This is not inconsistent with other professions, but still not laudable. In the business fields of science, technology, engineering and math, 74% of employees are male and 73% are White. Asian Americans account for 16% of the field, but other minorities make up less than 5%.[9] In the US Fish and Wildlife Service male employees outnumber females by more than 3:1, and in the US Forest Service the ratio is better, but men still are around 30% more common than women[10] — this despite the fact that women have outnumbered men in natural resource degree fields since 2000.[9] Time will tell whether these women are, in fact, being employed in meaningful occupations.

In many, many ways, conservation of natural resources in the United States and Canada has come a long way. We haven't solved all the problems, and new ones will continue to appear. However, if we maintain a strong workforce of dedicated, professionally trained technicians, managers, administrators and scientists, we can expect that the profession will only get stronger and better.

REFERENCES

1. PewResearchCenter. Most Americans believe climate change is real, but fewer see it as a threat. <http://www.pewresearch.org/fact-tank/2013/06/27/most-americans-believe-climate-change-is-real-but-fewer-see-it-as-a-threat/>; 2013.
2. Skeptical Science. Global warming vs climate change. <http://skepticalscience.com/climate-change-global-warming.htm>; 2013.
3. Boykoff MT, Boykoff J. Balances as bias: global warming and the US prestige press. *Global Environmental Change*. 2004;14:125–136.
4. Wake DB, Vredenburg VT. Are we in the midst of the sixth mass extinction? A view from the world of amphibians. *Proc Natl Acad Sci USA*. 2008;105:11466–11473.
5. International Union for the Conservation of Nature. Red List. <http://www.iucnredlist.org/documents/summarystatistics/2013_1_RL_Stats_Table_4b.pdf>; 2013.
6. US Geological Survey. The water cycle: freshwater storage. <http://ga.water.usgs.gov/edu/watercyclefreshstorage.html>; 2013.
7. Wildman RA, Forde NA. Management of water shortage in the Colorado River Basin: Evaluating current policy and the viability of interstate water trading. *J Am Water Res*. 2012;48:411–422.
8. Davis JM, Baxter CV, Rosi-Marshall EJ, et al. Anticipating stream ecosystem responses to climate change: toward predictions that incorporate effects via land–water linkages. *Ecosystems*. 2013;16:909–922.
9. Committee on Equal Opportunities in Science and Engineering. 2007–2008 Biennial Report to Congress. National Science Foundation; 2009.
10. Lopez R, Brown CH. Why diversity matters. Broadening our reach will sustain natural resources. *Wildlife Professional*. 2011;5:20–27.

Appendices

State	Agency Title	Wildlife	Fisheries	Boating	Parks	Forestry	Coastal & Marine	Soil & Water	Law	Mining
Alabama	Dept Conserv. & Natural Resources	x	x	x	x		x	x	x	
	Forestry Commission					x	x			
	Surface Mining Commission									x
Alaska	Dept Natural Resources	x			x	x		x		x
Arizona	Dept Fish & Game	x	x	x						
	Land Department					x		x	x	x
	Game & Fish Dept	x	x	x					x	
	State Parks				x					
Arkansas	Natural Resources Commission							x		
	Game & Fish Commission	x	x	x					x	
	Forestry Commission					x				
	Dept Parks and Tourism				x					
	Dept Environmental Quality									x
California	Natural Resources Agency	x	x	x	x	x	x	x	x	x
Colorado	Dept Natural Resources	x	x	x	x	x	x	x	x	x
Connecticut	Dept Energy & Environ. Protection	x	x	x	x	x	x	x	x	x

(Continued)

APPENDIX A: (Continued)

State	Agency Title	Wildlife	Fisheries	Boating	Parks	Forestry	Coastal & Marine	Soil & Water	Law	Mining
Delaware	Dept Natural Resources & Environ. Control	x	x	x	x		x	x	x	x
	Agriculture					x				
Florida	Fish & Wildlife Conserv. Comm.	x	x	x			x		x	
	Forest Service					x				
	Park Service				x					
	Dept Environ. Protection							x		x
Georgia	Dept Natural Resources	x	x	x	x		x	x	x	x
	Forestry Commission					x				
Hawaii	Dept Land & Natural Resources	x	x	x	x	x	x	x	x	x
Idaho	Dept Fish & Game	x	x	x						
	Dept Lands					x		x	x	x
	Parks & Recreation				x					
Illinois	Dept Natural Resources	x	x	x	x	x		x	x	x
Indiana	Dept Natural Resources	x	x	x	x	x		x	x	x
Iowa	Dept Natural Resources	x	x	x	x	x		x	x	x

(Continued)

APPENDIX A: (Continued)

State	Agency Title	Wildlife	Fisheries	Boating	Parks	Forestry	Coastal & Marine	Soil & Water	Law	Mining
Kansas	Dept Wildlife, Parks & Tourism	x	x	x	x	x		x	x	
	Dept Health & Environment									x
Kentucky	Dept Natural Resources					x		x		x
	Dept Fish & Wildlife Resources	x	x	x		x			x	
	State Parks				x					
Louisiana	Dept Natural Resources						x			x
	Dept Wildlife & Fisheries	x	x	x					x	
	Dept Agriculture and Forestry					x		x		
	Dept Culture, Rec. & Tourism				x					
Maine	Dept Conservation					x	x	x		x
	Dept Inland Fisheries & Wildlife	x	x	x	x				x	
Maryland	Dept Natural Resources	x	x	x	x	x	x	x	x	x
Massachusetts	Office Energy & Environ. Affairs	x	x	x	x	x	x	x	x	x
Michigan	Dept Natural Resources	x	x	x	x	x		x	x	
	Agric. & Rural Development									
	Dept Environmental Quality									x

(Continued)

APPENDIX A: (Continued)

State	Agency Title	Wildlife	Fisheries	Boating	Parks	Forestry	Coastal & Marine	Soil & Water	Law	Mining
Minnesota	Dept Natural Resources	x	x	x	x	x		x	x	
Mississippi	Dept Wildlife, Fisheries & Parks	x	x		x			x	x	
	Forestry Commission					x				
	Dept Environmental Quality									x
	Dept Marine Resources						x			
Missouri	Dept Conservation	x	x	x		x			x	
	Dept Natural Resources				x			x		x
Montana	Dept Natural Resources & Conserv.					x		x		x
	Fish, Wildlife & Parks	x	x	x	x				x	
	Dept Environmental Quality									x
Nebraska	Game & Parks Commission	x	x	x	x				x	
	Forest Service					x				
	Dept Natural Resources							x		
	Dept Environmental Quality									x

(Continued)

State	Agency Title	Wildlife	Fisheries	Boating	Parks	Forestry	Coastal & Marine	Soil & Water	Law	Mining
Nevada	Dept Wildlife	x	x	x					x	
	Dept Conserv. & Natural Resources				x	x		x		x
New Hampshire	Fish & Game Dept	x	x	x			x		x	
	Dept Resources & Economic Development				x	x		x		
	Geologic Survey									x
New Jersey	Dept Environmental Protection	x	x	x	x	x	x	x	x	x
New Mexico	Minerals & Natural Resources Dept	x	x	x	x	x		x	x	
New York	Dept Environmental Conserv.	x	x	x	x	x		x	x	x
North Carolina	Dept Environ. & Natural Resources	x	x	x	x		x	x	x	x
	Dept Agric. & Consumer Services					x				
North Dakota	Game & Fish Dept	x	x	x	x				x	
	Parks & Recreation				x			x		
	Forest Service					x				
	Industrial Commission									x

(Continued)

State	Agency Title	Wildlife	Fisheries	Boating	Parks	Forestry	Coastal & Marine	Soil & Water	Law	Mining
Ohio	Dept Natural Resources	x	x	x	x	x		x	x	x
Oklahoma	Dept Wildlife Conservation	x	x	x				x	x	
	Dept Agric., Food & Forestry					x				
	Tourism & Recreation Dept				x					
	Dept Mines									x
	Conservation Commission							x		
Oregon	Dept Fish & Wildlife	x	x	x					x	
	Dept Forestry					x				
	Parks & Recreation Dept				x					
	Dept State Lands									x
	Dept Agriculture							x		
	Dept Land Conserv. & Develop.						x			
Pennsylvania	Conserv. & Natural Resources				x	x		x	x	x
	Fish & Boat Commission		x	x						
	Game Commission	x							x	
	Minerals Resources Mgmt									x

(Continued)

State	Agency Title	Wildlife	Fisheries	Boating	Parks	Forestry	Coastal & Marine	Soil & Water	Law	Mining
Rhode Island	Dept Environmental Mgmt	x	x	x	x	x	x	x	x	x
South Carolina	Dept Natural Resources	x	x	x			x	x	x	x
	Dept Parks, Rec. & Tourism				x					
	Forestry Commission					x				
	Dept Health & Environ. Control									x
South Dakota	Fish, Game & Parks	x	x	x	x				x	
	Environ. & Natural Resources							x		x
	Dept Agriculture					x				
Tennessee	Wildlife Resource Agency	x	x	x					x	
	Dept Agriculture					x				
	Dept Environment & Conserv.				x			x		x
Texas	Parks & Wildlife	x	x	x	x		x	x	x	
	Commission Environ. Quality							x		x
	Forest Service					x				
Utah	Dept Natural Resources	x	x	x	x	x		x	x	x
Vermont	Agency Natural Resources	x	x	x	x	x		x	x	x

(Continued)

APPENDIX A: (Continued)

State	Agency Title	Wildlife	Fisheries	Boating	Parks	Forestry	Coastal & Marine	Soil & Water	Law	Mining
Virginia	Dept Environmental Quality	x	x	x	x	x	x	x	x	x
Washington	Dept Natural Resources					x	x	x	x	x
	Dept Fish & Wildlife	x	x	x						
	State Parks & Recreation Commission				x					
West Virginia	Dept Commerce Div. Natural Resources	x	x	x	x	x		x	x	x
Wisconsin	Dept Natural Resources	x	x	x	x	x		x	x	x
Wyoming	Game and Fish Dept	x	x	x						
	State Parks, Historic Sites & Trails				x				x	
	State Lands & Investments					x		x	x	x

APPENDIX B: Provincial Agencies and Their Responsibilities.

Province	Agency Title	Wildlife	Fisheries	Parks	Forestry	Coastal & Marine	Soil & Water	Law	Mining
Alberta	Environ. & Sustain. Resource Development	×	×	×	×	×	×	×	
	Energy								×
British Columbia	Forests, Lands & Natural Resources Ops	×	×	×	×	×	×	×	
	Mining Association of BC								×
Manitoba	Conserv. & Water Stewardship	×	×	×	×		×	×	
	Innovation, Energy and Mines								×
New Brunswick	Natural Resources	×	×		×			×	
	Agric., Aquaculture & Fisheries					×			
	Tourism, Heritage & Culture			×					
	Energy & Mines								×
	Dept Environment					×	×		
Newfoundland and Labrador	Environment & Conservation	×		×				×	
	Natural Resources				×		×		×
	Fish & Aquaculture		×			×			

(Continued)

APPENDIX B: (Continued)

Province	Agency Title	Wildlife	Fisheries	Parks	Forestry	Coastal & Marine	Soil & Water	Law	Mining
Nova Scotia	Environment						x		
	Natural Resources	x		x	x			x	x
	Fish & Aquaculture		x			x			
Ontario	Natural Resources	x	x	x	x		x	x	x
Prince Edward Island	Agriculture & Forestry	x	x	x	x				
	Tourism & Culture			x					
	Fish, Aquaculture & Rural Develop.					x			
	Energy & Minerals								x
	Environ., Labor & Justice						x		
Quebec	Ressources Naturelles	x	x		x		x	x	x
	Parcs Quebec			x					
Saskatchewan	Environment	x	x		x		x	x	
	Parks, Culture & Sport			x					
	Energy & Resources								x

Index

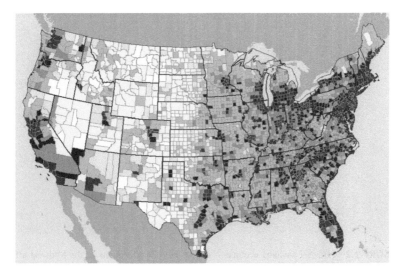

FIGURE I.5 Distribution of human population within the United States (the darker the color, the greater the population). While there are areas in the west that have low populations, coastal areas are crowded.

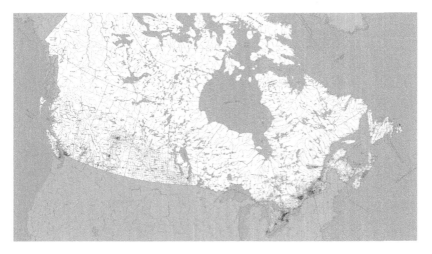

FIGURE I.6 Distribution of human population within Canada. Note that most of the people are clustered in the southern portion of the nation. *Credit: Statistics Canada.*

FIGURE 2.2 Midwest savanna ecotone — one of the several regions in the Midwest affected by fire and climate. *Credit: World Wildlife Fund, Ecological Encyclopedia.*

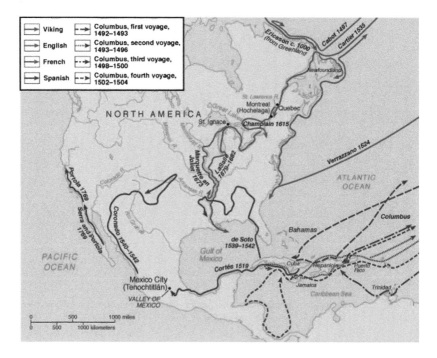

FIGURE 2.4 Early exploration routes of Columbus, the Spanish and the French.

Percent

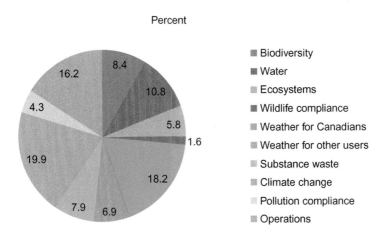

- Biodiversity
- Water
- Ecosystems
- Wildlife compliance
- Weather for Canadians
- Weather for other users
- Substance waste
- Climate change
- Pollution compliance
- Operations

FIGURE 5.3 The annual budget in terms of percent of total budget for Environment Canada. *Data: Treasury Board of Canada Secretariat.*

Percent

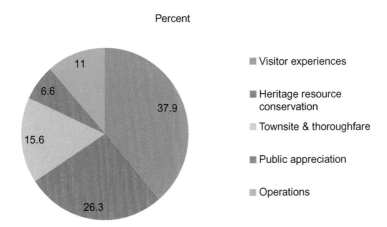

- Visitor experiences
- Heritage resource conservation
- Townsite & thoroughfare
- Public appreciation
- Operations

FIGURE 5.6 The annual budget in terms of percent total budget for Natural Resources Canada. *Data: Treasury Board of Canada Secretariat.*

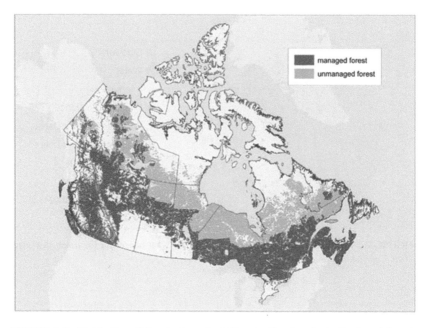

FIGURE 5.8 Distribution of forests throughout Canada. Due largely to the boreal forests, Canada is among the top 4 countries in the world in terms of amount of forest, much of which is unbroken. *Credit: Natural Resources Canada.*

Percent

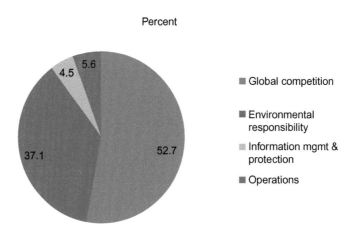

5.6

4.5

37.1

52.7

■ Global competition

■ Environmental responsibility

▧ Information mgmt & protection

■ Operations

FIGURE 5.9 Annual budget for Natural Resources Canada in terms of percent total budget. *Data: Treasury Board of Canada Secretariat.*

Percent

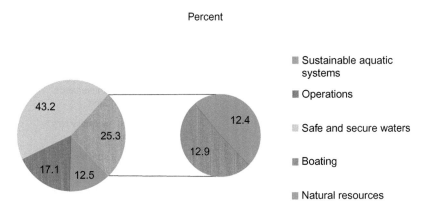

Sustainable aquatic systems

Operations

Safe and secure waters

Boating

Natural resources

FIGURE 5.11   Annual budget for Fisheries and Oceans in terms of percent total budget. *Data: Treasury Board of Canada Secretariat.*

Percent

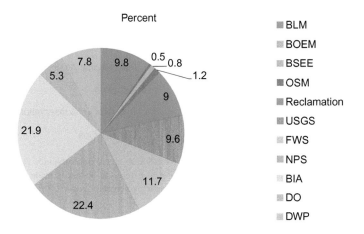

BLM

BOEM

BSEE

OSM

Reclamation

USGS

FWS

NPS

BIA

DO

DWP

FIGURE 6.2   Annual Budget for the Department of Interior by Agency. *Credit: US Department of the Interior.*

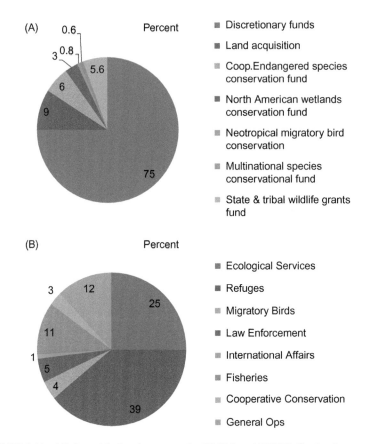

FIGURE 6.11 (A) Annual budget by percent for US Fish and Wildlife Service Programs. All but the discretionary funds portion are obligated even before the budget is made. (B) US Fish and Wildlife Service distribution of discretionary funds by percent. *Credit: US Fish and Wildlife Service.*

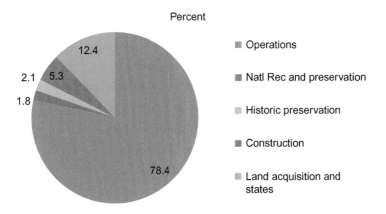

Percent

■ Operations

■ Natl Rec and preservation

■ Historic preservation

■ Construction

■ Land acquisition and states

FIGURE 6.17 Distribution of National Park Service Funds by Program. The biggest portion is Operations, which includes running the parks. *Credit: US National Park Service.*

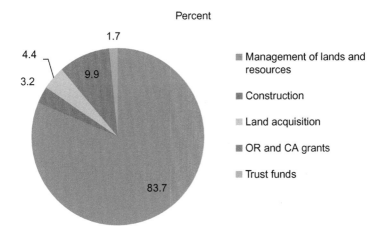

Percent

■ Management of lands and resources

■ Construction

■ Land acquisition

■ OR and CA grants

■ Trust funds

FIGURE 6.18 Distribution of Bureau of Land Management funds by program. *Credit: Bureau of Land Management.*

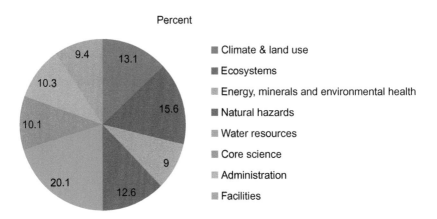

Percent

■ Climate & land use

■ Ecosystems

▨ Energy, minerals and environmental health

■ Natural hazards

▨ Water resources

■ Core science

▨ Administration

▨ Facilities

FIGURE 6.22 Annual budget by program for the US Geological Survey. *Information from US Geological Survey.*

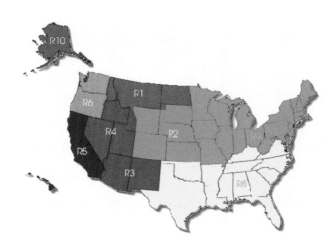

FIGURE 7.6 The nine US Forest Service regions. *Credit: US Forest Service.*

Forest service

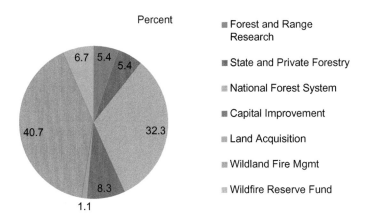

Percent

- Forest and Range Research
- State and Private Forestry
- National Forest System
- Capital Improvement
- Land Acquisition
- Wildland Fire Mgmt
- Wildfire Reserve Fund

FIGURE 7.8 The budget of the US Forest Service based on percent of total budget. Information from US Forest Service.

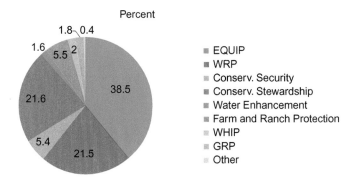

Percent

- EQUIP
- WRP
- Conserv. Security
- Conserv. Stewardship
- Water Enhancement
- Farm and Ranch Protection
- WHIP
- GRP
- Other

FIGURE 7.12 The annual mandatory budget for NRCS. The total budget for NRCS approximates $4.5 billion, but the mandatory portion, seen here, is around 80% of that, or approximately $3.6 billion. *Information from US Department of Agriculture.*

Billions of Dollars

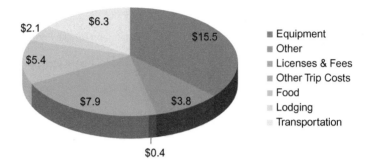

FIGURE 10.4 Expenditures in billions of dollars spent on fishing during 2011. The amount of money spent on equipment was by far the biggest cost. *Data courtesy of US Fish and Wildlife Service.*

Billions of Dollars

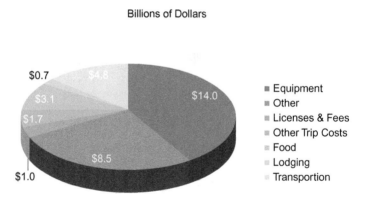

FIGURE 10.5 Expenditures in billions of dollars spent on hunting during 2011. Again, equipment was the biggest cost. *Data courtesy of US Fish and Wildlife Service.*

Billions of Dollars

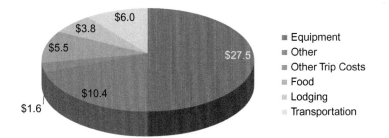

Equipment
Other
Other Trip Costs
Food
Lodging
Transportation

FIGURE 10.7 Total expenditures for people watching wildlife in the United States during 2011. As with hunting and fishing, the greatest cost was equipment. The *per capita* cost for watching wildlife is smaller than the other activities, but the much larger number of people who watch than fish or hunt makes up for the difference in total costs. *Data courtesy of the US Fish and Wildlife Service.*

Millions of dollars

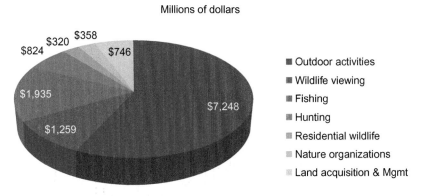

Outdoor activities
Wildlife viewing
Fishing
Hunting
Residential wildlife
Nature organizations
Land acquisition & Mgmt

FIGURE 10.11 Total expenditures on fishing, hunting and wildlife observation in Canada during 1996. *Data courtesy of Statistics Canada.*

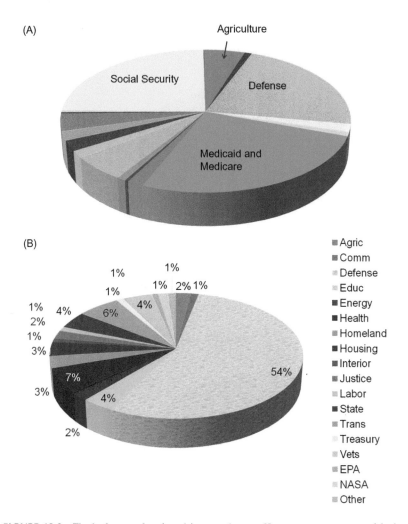

(A)

Agriculture

Social Security

Defense

Medicaid and
Medicare

(B)

■ Agric
■ Comm
▩ Defense
▨ Educ
■ Energy
■ Health
■ Homeland
■ Housing
■ Interior
■ Justice
▨ Labor
■ State
▨ Trans
▨ Treasury
▨ Vets
▦ EPA
▨ NASA
▨ Other

1% 1%
1% 1% 2% 1%
1% 4% 4%
2% 6%
1%
3%
7%
3%
4%
2%
54%

FIGURE 12.3   The budget can be viewed in several ways. Here we see a recent total budget for the United States (A) and the portion of the budget that is only discretionary (B). Social Security, Medicare, Farm Support and other programs take up a large portion of the budget, but the Defense Department's budget is huge in either category.